THE ASTROPHOTOGRAPHER'S GUIDEBOOK

THE ASTROPHOTOGRAPHER'S GUIDEBOOK

ANTOINE & DALIA GRELIN
GALACTIC-HUNTER.COM

CONTENTS

FOREWORD	6
HOW TO READ THIS BOOK	8

SPRING 10

MARKARIAN'S CHAIN	12
M104	14
NGC 2237	16
M101	18
M66 GROUP	20
M63	22
M81 & M82	24
M97	26
NGC 4631	28
M51	30
IC 443	32
NGC 4565	34
M64	36
M106	38
NGC 2359	40
SUMMARY	42

SUMMER 44

M8	46
NGC 6888	48
M20	50
M27	52
M17	54
M57	56
M24	58
M16	60
RHO OPHIUCHI	62
NGC 7000 & IC 5070	64
M11	66
NGC 7380	68
M13	70
NGC 6960	72
M75	74
SUMMARY	76

WILLIAM H. MILLER
HIGH STYLE ON THE HIGH SEAS
PASSENGER SHIPS INTERIORS

FALL	78	WINTER	112
M31	80	M1	114
M15	82	M35	116
IC 1805	84	NGC 1499	118
IC 1848	86	M45	120
M33	88	M44	122
NGC 7293	90	M95	124
IC 1396	92	M96	126
HCG 92	94	M37	128
NGC 7635	96	M42	130
NGC 7023	98	IC 434	132
IC 5146	100	IC 2118	134
M74	102	M78	136
M92	104	BARNARD'S LOOP	138
IC 405	106	NGC 869 & NGC 884	140
NGC 281	108	NGC 2264	142
SUMMARY	110	SUMMARY	144

SUMMARY	146
END WORD	152
CREDITS	153

FOREWORD

We are Antoine and Dalia Grelin, a couple who have been enjoying the dark skies of the Nevada desert for several years. Our little stargazing outings quickly transformed into a hobby for astrophotography. We began a quest to capture all of the Messier objects which turned into Galactic Hunter, a YouTube channel where we make videos about photographing Deep Sky Objects.

Our videos demonstrate how to photograph celestial bodies from A to Z. Our process often begins from our home in Las Vegas and the preparations, as well as precautions we take before adventuring out into the darkness. We also show the step-by-step instructions that goes in to processing each photo from our point of view.

Using our experience with DSLR Astrophotography, we decided to make a simple in-depth guide of the best targets for amateur astrophotographers. If you would like to see more of our images, tutorials, or our videos, visit galactic-hunter.com.

Capturing the Horse
01/28/2018
5pm - 1am

TONIGHT'S TARGET
IC 434

CONSTELLATION
Orion

MEMORIES OF THE NIGHT

After a very long break, thanks to the clouds and the wind, I was able to finally go back to my imaging spot.

Tonight, the weather was perfect, no clouds, no wind, no humidity, and the temperature felt pretty nice!

The plan was to capture the HorseHead and Flame nebulae, and after about 4 hours of imaging... I got it pretty good!

While the camera and telescope were busy, I watched a scary movie in the car (note to self: Do not watch scary movies, alone in the middle of nowhere, at night). It was ok. Then, surprisingly, a couple, Mark and Melanie, arrived at the location and installed their own telescope to do some visual!
It was awesome because this was the first time I met other astronomers, and we had an amazing time! We saw M42, M78, and the Christmas Tree cluster using their 10" telescope.

BEST IMAGE OF THE NIGHT

Inverted Mask during Processing

ACQUISITION DETAILS
- Total exposure: 1 hours
- 10 lights of 6 minutes each
- Calibrated with 15 darks, 15 bias

TIPS FOR FUTURE
- Install all the gear before sunset or it gets too dark.
- Don't watch scary movies alone in the desert...

DIFFICULTY
★★☆

For a better experience on your journey to capture the greatest images of the night sky, we recommend pairing this guide with "The Astrophotographer's Journal", a notebook to record each outing and track your progress in astrophotography.

This book contains the 60 best astrophotography targets of the year and was written with the goal of allowing any amateur astrophotographer, beginner or intermediate, to have a helpful guide on their hunt through deep space. In order to keep this book convenient and easy to follow, we made its design simple and straight to the point.

The targets chosen in the chapters to follow are, in our opinion, the best Deep Sky Objects to photograph before capturing more complex images. Our experience has taught us to learn from our mistakes and to understand that failures are stepping stones to success. We hope you will learn a lot about astronomy and astrophotography while developing your skills when crossing off each target as the seasons go by.

Clear Skies,
Galactic Hunter

HOW TO READ THIS BOOK

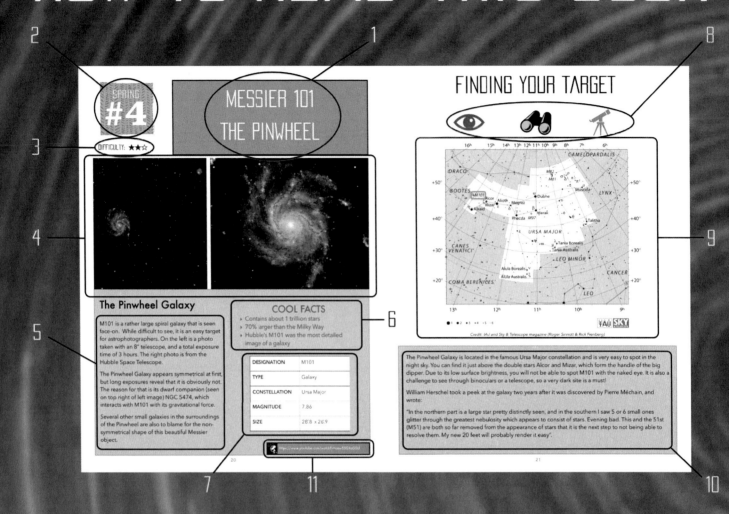

Each double page in this book featuring a target follows the same format, as seen in the image above.

Upon reading, there are helpful key facts about the Deep Sky Object you are planning to image, such as its designation, how to find it in the night sky, advice on imaging, a statistics table, and more!

A few targets will have the Galactic Hunter logo and a link on the bottom right of the page. This link will direct you to an episode that will help you capture that specific target. This can be useful as you will hear us explain more in depth about the object and perhaps determine if said target is a good choice for you to photograph.

1 - DESIGNATION & NAME OF TARGET

The object's astronomical naming, or designation is displayed first, followed by its commonly used name. Most Deep Sky Objects in this book have common names (such as the Andromeda Galaxy, or the Heart Nebula), but a few do not. In that case, we describe the object and its location, for example: "Double Cluster in Perseus".

2 - BEST SEASON TO PHOTOGRAPH

The square on the top left indicates the object's number in our guide, as well as the best season to observe and photograph it. The square is also color coded, so that Spring is green, Summer is yellow, Fall is brown, and Winter is blue.

3 - DIFFICULTY LEVEL

The Difficulty Level is rated using three stars. Do not let a Difficulty Rating of three stars scare you! This guide is dedicated to the best and easiest targets of the year, so an object with a three star Difficulty Level is considered intermediate at most.

4 - EXAMPLE PHOTOS

Two images are attached to each target page. Occasionally, it is a comparison between the Hubble Space Telescope and what an amateur astrophotographer can expect, and other times it shows the difference between a DSLR camera image and a CCD's.

5 - QUICK DESCRIPTION AND ADVICE

A short description of the target followed by some personal advice to capture it.

6 - COOL FACTS

A few interesting facts about the object that may be surprising!

7 - STATISTICS TABLE

Table that includes the target's designation, its type, the constellation in which it is located, its magnitude, and size.

8 - VISIBILITY ICONS

Three icons to help you quickly determine if the target is visible to the naked eye, through binoculars, or through a telescope only. A green icon means that it is visible and a red icon means that it is not.

9 - LOCATION MAP

A map of the night sky focused on the target's location within its constellation. The target's position is marked by a rounded, orange square.

10 - VISIBILITY & LOCATION DESCRIPTION

A short paragraph to help you locate your target, most of the time by star hopping, as well as our notes about what the object looks like visually.

11 - LINK TO OUR CAPTURE

Many targets in this book have a link at the bottom of the page. Going to galactic-hunter.com and adding the URL at the end will take you to our complete blog post about photographing the object. It is full of tips and information, so have a look!

THE 15 BEST TARGETS of...

SPR

ING

SPRING IS FAMOUS FOR BEING THE GALAXY SEASON! OUT OF THE 15 BEST TARGETS FOR SPRING, YOU WILL FIND THAT 11 OF THEM ARE GALAXIES AND THE REMAINING 4 ARE BEAUTIFUL NEBULAE.

THERE ARE NO CLUSTERS LISTED DURING THE SPRING TIME AS THE BEST BEGINNER ONES ARE RESERVED FOR THE NEXT SEASON.

SPRING #1

DIFFICULTY: ★★★

VIRGO CLUSTER MARKARIAN'S CHAIN

Markarian's Chain

Let's start off with not one, not two, but a bunch of galaxies, all forming a chain!

Markarian's Chain (right) is a stretch of more than eight galaxies located in the heart of the famous Virgo Cluster. Astrophysicist Benjamin Markarian named it after discovering that at least seven of them appeared to be moving coherently.

Photographing the chain is easy, as you can accomplish this with any small telescope, or even a camera lens and star tracker. Be sure to frame your image in a way that the two main interacting galaxies, Markarian's Eyes (left) are included!

Because of their size and faint details, be sure to spend at least four hours on this cluster of galaxies or you might end up with blobs of light.

COOL FACTS

- Named after Armenian astrophysicist Benjamin Markarian
- Stretches over a distance of 20 full moons
- Contains several Messier Objects

DESIGNATION	Virgo Cluster
TYPE	Chain of Galaxies
CONSTELLATION	Virgo
MAGNITUDE	9.0 - 15.0
SIZE	1°+

- Suggested minimum focal length: 85mm
- Ideal focal length: 350mm

FINDING YOUR TARGET

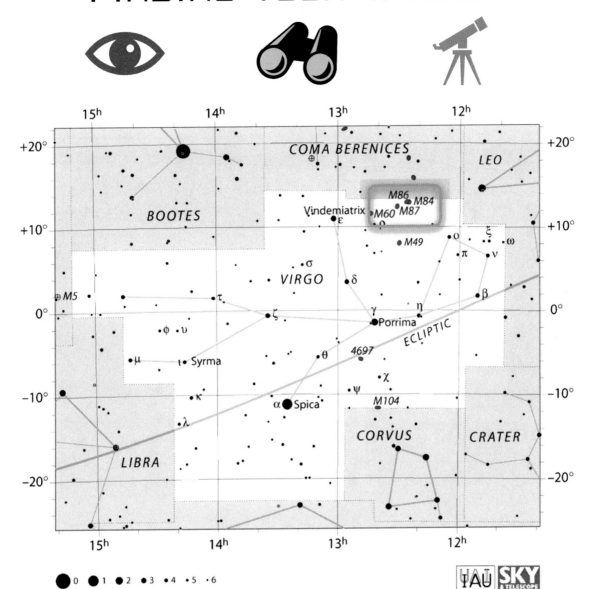

The brightest members of Markarian's Chain can be seen through binoculars and small telescopes, but you will need a higher power telescope to be able to see the fainter galaxies.

The Virgo cluster is one of the best sights seen with binoculars in the night sky during Spring. From a dark site, you will be able to not only see the members of the chain, but also some additional bright galaxies around, like Messier 87!

Markarian's Chain can be found on the outer edge of the Virgo constellation, it lies right between the stars Denebola, in Leo, and Vindemiatrix, in Virgo. Draw an imaginary line between those two bright stars and at the center is your target!

Whether you are taking pictures or observing through an eyepiece, challenge yourself to match each galaxy with its proper designation!

/post/markarians-chain

SPRING #2

DIFFICULTY: ★★☆

MESSIER 104
THE SOMBRERO

The Sombrero Galaxy

Messier 104 appears as a very small and thin object, but it holds the biggest supermassive black hole ever recorded in any nearby galaxy. Despite its small size, it is a popular target among amateur astrophotographers because of its beautiful dust lanes that cross in front of its luminous center.

The photo on the left was taken with an 8" telescope, compiled of 60 frames at 3 minutes each for a total of 3 hours.

The real challenge lies in the photo processing, as you must be sure to not make the core of the galaxy too bright otherwise you will end up with an image that looks overexposed.

COOL FACTS

- Getting farther from us at 1,000 km/s
- Has a supermassive black hole with a mass of more than 1 billion suns
- Brightest galaxy in a 10 megaparsec radius

DESIGNATION	M104
TYPE	Galaxy
CONSTELLATION	Virgo
MAGNITUDE	8.98
SIZE	8'.7 x 3'.5

- Suggested minimum focal length: 650mm
- Ideal focal length: 1200mm

FINDING YOUR TARGET

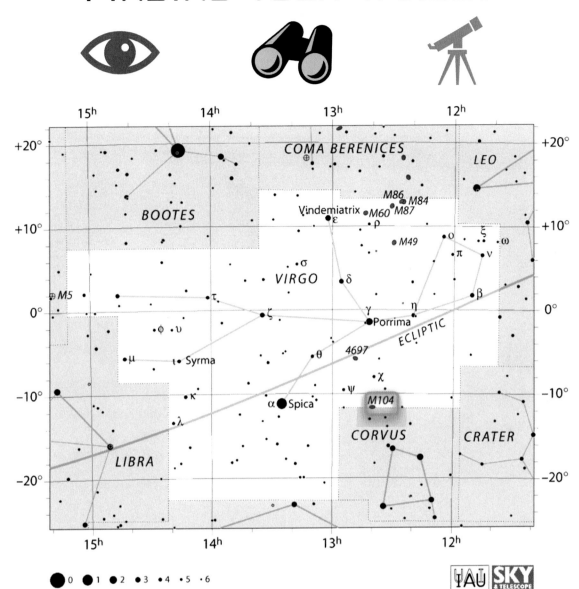

The Sombrero galaxy can be seen from extremely dark sites with binoculars or telescopes, but is not an impressive target to look at. With a long focal length telescope, you may be able to distinguish the dark dust lane in front of the galaxy's bright bulge.

To locate it, look for the constellation of Virgo and find its brightest star, Spica. M104 can be spotted just 11.5 degrees west of the bright star.

Another nice feature to look at if you're photographing or observing M104 is the nearby asterism known as "The Jaws". This colorful grouping of stars can be seen near the right edge of our first image attached.

/blog/m104-the-

SPRING #3

DIFFICULTY: ★☆☆

NGC 2237
THE ROSETTE

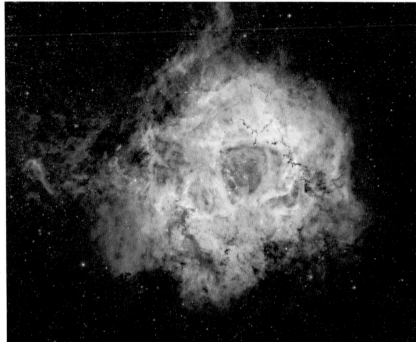

The Rosette Nebula

The Rosette is one of the easiest nebulae to capture and unlike M104, it is huge!

NGC 2237 is best captured using a small telescope as you will be able to also include the outer gas. The Rosette contains an open cluster of stars, NGC 2244, which can be seen around the center of both images above.

The nebula will look almost completely red if using a camera without filters. It is very rich in Hydrogen alpha and Oxygen III gas, and so is a perfect target to re-visit after purchasing a duo band filter or monochrome camera. The picture on the left shows a true-color image of the target, while the one on the right shows a Hubble Palette combination using narrowband filters.

COOL FACTS

- Discovered in 5 different parts over time
- Also called "The Skull"
- One of the most massive emission nebulae

DESIGNATION	NGC 2237
TYPE	Nebula
CONSTELLATION	Monoceros
MAGNITUDE	9.0
SIZE	1.3°

- Suggested minimum focal length: 50mm
- Ideal focal length: 300mm

FINDING YOUR TARGET

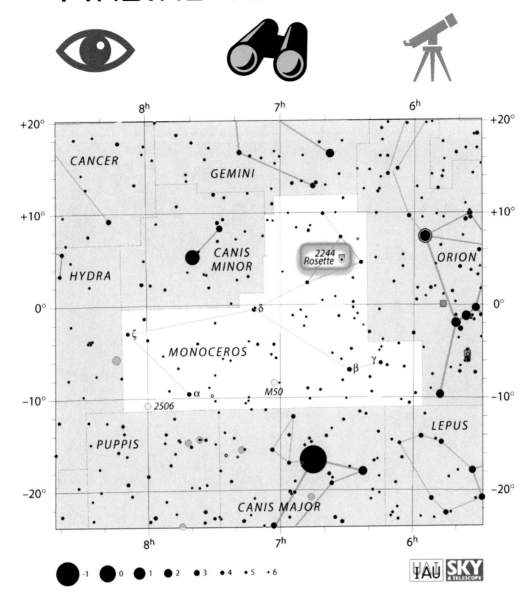

NGC 2237 can be seen in the constellation of Monoceros, the Unicorn.

The cluster of stars in its center can be easily seen through binoculars or a telescope. The nebulosity itself can also be seen slightly through binoculars if far from light pollution. A small telescope is recommended to see the Rosette clearly, as it is harder to spot when using a larger instrument.

Not far from the Rosette Nebula is another large and bright object, the Christmas Tree and Cone Nebula. This target, along with the Rosette, can be captured in one frame if imaged without a telescope.

If this is something you wish to attempt, you can try using a full-frame DSLR or mirrorless camera paired with a 135mm lens and aim between the two objects. If your camera has a cropped-sensor, try a lens with a focal length of around 85mm instead. It is of course recommended to use a star tracker for this.

/post/ngc-2244

SPRING #4

DIFFICULTY: ★★☆

MESSIER 101
THE PINWHEEL

The Pinwheel Galaxy

M101 is a large spiral galaxy that is seen head-on. While difficult to see, it is an easy target for astrophotographers. The photo on the left was taken with an 8" Newtonian telescope with a total exposure time of 3 hours. The photo on the right is from the Hubble Space Telescope.

The Pinwheel Galaxy appears symmetrical at first, but long exposures reveal that it is not. This is due to its dwarf companion (seen on top right of the left image) NGC 5474, which is pulling M101 with its gravitational force.

Several small galaxies surrounding the Pinwheel Galaxy are also to blame for the asymmetrical shape of this beautiful Messier object.

COOL FACTS

- Contains approximately 1 trillion stars
- 70% larger than the Milky Way
- Hubble's M101 was the most detailed image of a galaxy at the time it was taken

DESIGNATION	M101
TYPE	Galaxy
CONSTELLATION	Ursa Major
MAGNITUDE	7.86
SIZE	28'.8 x 26'.9

- Suggested minimum focal length: 250mm
- Ideal focal length: 1000mm

FINDING YOUR TARGET

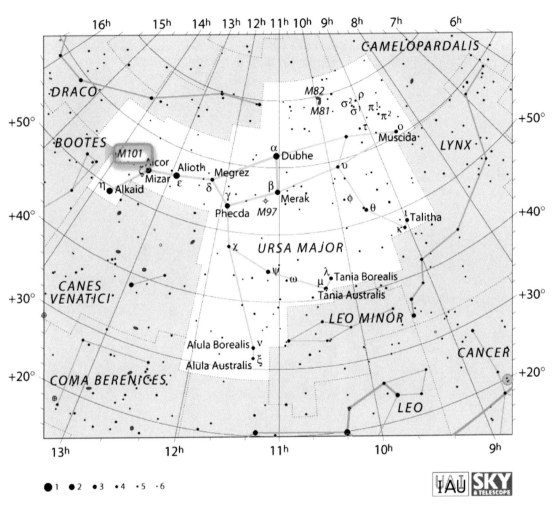

The Pinwheel Galaxy is located in the famous Ursa Major constellation and is very easy to spot in the night sky. You can find it just above the double stars Alcor and Mizar, which form the handle of the big dipper. Due to its low surface brightness, you will not be able to spot M101 with the naked eye. It is also a challenge to see through binoculars or a telescope, so viewing this target from a dark site is a must!

William Herschel took a peek at the galaxy two years after it was discovered by Pierre Méchain, and wrote:

"In the northern part is a large star pretty distinctly seen, and in the southern I saw 5 or 6 small ones glitter through the greatest nebulosity which appears to consist of stars. Evening bad. This and the 51st (M51) are both so far removed from the appearance of stars that it is the next step to not being able to resolve them. My new 20 feet will probably render it easy".

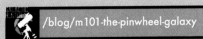

SPRING #5

DIFFICULTY: ★★★

M66 GROUP
LEO TRIPLET

Leo Triplet

The Leo Triplet is composed of two Messier objects, M65 and M66 (left) as well as NGC 3628, which is also called the Hamburger galaxy due to its shape (bottom left on the right image).

The M66 Group gets its 3-star difficulty rating for a few reasons. First, make sure to center your telescope and angle your camera accordingly so that all three galaxies fit nicely in the frame. This becomes harder the larger the instrument is. The second reason is the processing. Ensure that all three targets have a similar brightness, saturation, and about the same amount of outer dust showing. Lastly, the Hamburger galaxy has a nice tidal tail that will only reveal itself with enough data and great processing skills!

COOL FACTS

▸ May be part of a bigger group
▸ Messier 66 is the brightest of the three
▸ All deformed by each other's gravity

| DESIGNATION | M65
M66
NGC 3628 |
|---|---|
| TYPE | Group of Galaxies |
| CONSTELLATION | Leo |
| MAGNITUDE | M65 - 10.25
M66 - 8.9
NGC 3628 - 10.2 |
| SIZE | M65 - 8'.7 x 2'.5
M66 - 9'.1 x 4'.2
NGC 3628 - 15' x 3'.6 |

FINDING YOUR TARGET

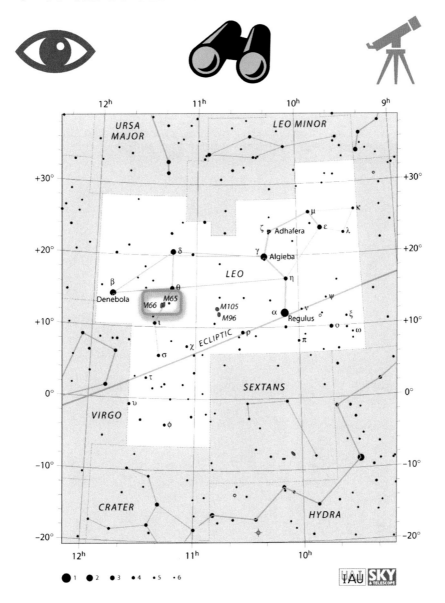

The Leo Triplet can be found in the constellation of Leo, not far from Ursa Major and Virgo.

To locate it, first find the star Chertan which makes up part of the back legs of the lion, then move downward towards Virgo. You are likely to land on at least one of the three galaxies, so missing the remaining two is nearly impossible.

In dark skies, far from any light pollution, all three galaxies can be observed through binoculars. An medium size telescope is ideal for viewing the group in one field of view.

- Suggested minimum focal length: 350mm
- Ideal focal length: 800mm

/post/leotriplet

SPRING #6

DIFFICULTY: ★★☆

MESSIER 63
THE SUNFLOWER

The Sunflower Galaxy

M63 is a spiral galaxy located about 37 million light-years away from Earth.

The Sunflower Galaxy got its name from its yellow core and the shape of its arms resembling a sunflower. Photographing this target is not difficult, but it received a 2-star rating due to its small size and the difference of brightness between the core and the arms. We recommend at least four hours of exposures on this galaxy, as the beautiful details in the arms, or petals, will not show if the exposure time is too short.

Processing the Sunflower is very exciting, as the galaxy will look somewhat dull until you adjust the colors and saturation to transform it into a beauty!

COOL FACTS

- Pierre Méchain's first discovery, in 1779
- Is a member of the M51 Group
- One of the first galaxies where a spiral structure was identified

DESIGNATION	M63
TYPE	Galaxy
CONSTELLATION	Canes Venatici
MAGNITUDE	9.3
SIZE	12'.6 x 7'.2

- Suggested minimum focal length: 600mm
- Ideal focal length: 1000mm

FINDING YOUR TARGET

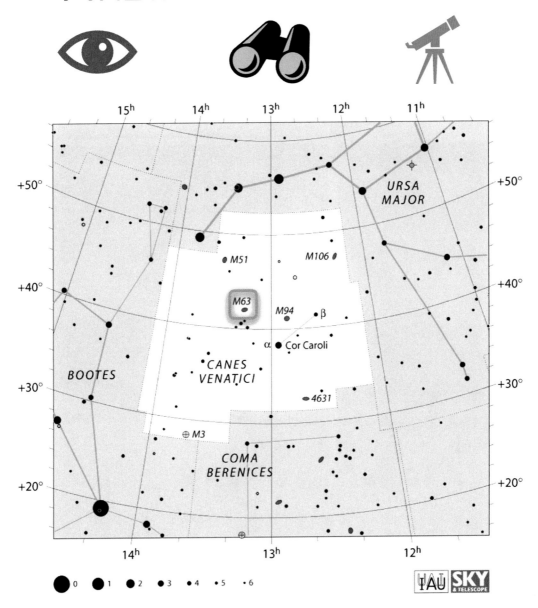

M63 lies in the constellation Canes Venatici, which is a rather faint constellation because none of its stars pop like the surrounding ones.

The easiest way to find its location is to start from Ursa Major. Locate the last star of the Big Dipper's handle, Alkaid, then begin moving towards the brightest star of the Canes Venatici constellation: Cor Caroli. You should spot your target about halfway between both stars.

The Sunflower galaxy is bright enough to be seen with binoculars and small telescopes, but you will only be able to spot a small gray smudge. Telescopes with a focal length of 800mm or more will reveal the bright core as well as some of the gases around it.

SPRING #7

DIFFICULTY: ★★☆

MESSIER 81 & 82
BODE'S & THE CIGAR

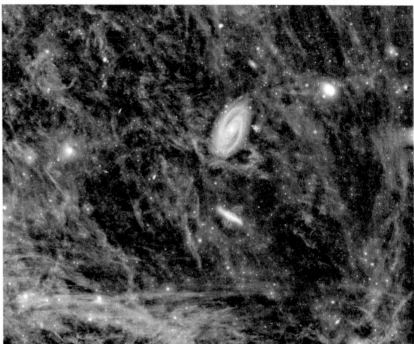

Bode's and the Cigar Galaxy

Spiral galaxy M81 and starburst galaxy M82 (left) are two magnificent neighboring galaxies, that are almost always photographed as a pair.

When taking a photo of the two, make sure to double check that both galaxies fit nicely in the frame of the camera. The photo on the right was taken with a small refractor telescope under very dark skies. It totals 20 hours of exposure time and shows the very faint IFN (Integrated Flux Nebula) all over the field of view. For your first attempt, do not attempt to get this gas and just focus on the two galaxies!

Processing-wise, it may be a challenge to make the red hues of the Cigar galaxy visible. Longer exposure times or the use of a Hydrogen Alpha filter will help to bring out the starburst region.

COOL FACTS

- The pair were discovered in 1774
- M82 is the closest starburst galaxy to Earth
- M81's tidal forces affect M82 and increases its star forming activity

DESIGNATION	M81 M82
TYPE	Galaxies
CONSTELLATION	Ursa Major
MAGNITUDE	M81 - 6.94 M82 - 8.41
SIZE	M81 - 26'.9 x 14'.1 M82 - 11'.2 x 4'.3

FINDING YOUR TARGET

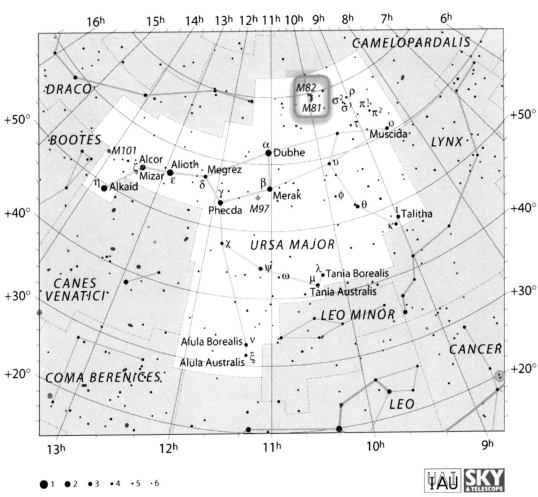

Both galaxies are visible through binoculars and telescopes, but not with the naked eye. Depending on the instrument, M81 will look like a blurry oval shape with a bright center, while M82 will appear as a thin line of light.

The pair are located in the constellation of the Big Dipper: Ursa Major. The easiest way to find them is to first spot the bright star Dubhe, which forms the top point of the Big Dipper pan, then travel about 10 degrees northwest to spot the two galaxies.

The IFN is of course impossible to see, being some of the faintest type of gasses in the universe.

- Suggested minimum focal length: 250mm
- Ideal focal length: 800mm to fit both, 1200mm individually

 /blog/m81-m82-bode-s-galaxy-the-cigar-galaxy

SPRING #8

DIFFICULTY: ★★★

MESSIER 97
THE OWL

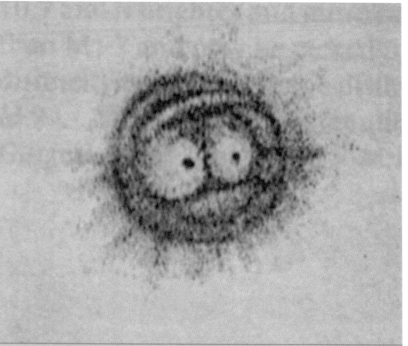

The Owl Nebula

The Owl nebula may be tiny, dim, and boring looking, but it is a popular target for amateur astrophotographers. The main reason being it lies close to another Messier object: The Surfboard galaxy, or M108. They are close enough together that both targets can be imaged in the same field of view! You can see the Owl top center (left image), and M108 in the middle left. The green blob near the bottom is actually a comet designated as 41P!

The Owl nebula gets a 3-star difficulty rating because of its size and that it is a dim target. We advise spending several hours of imaging if you expect to see details in your final image and not just a blue ball floating in space.

COOL FACTS

- Very similar to the Southern Owl Nebula
- Got its name after William Parsons' drawing in 1848 (right)
- Appears brighter visually than in photos

DESIGNATION	M97
TYPE	Nebula
CONSTELLATION	Ursa Major
MAGNITUDE	9.9
SIZE	3'.4 x 3'.3

FINDING YOUR TARGET

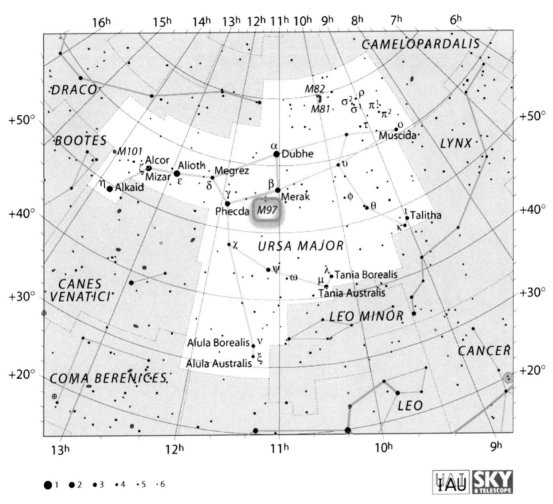

Because of its small size and faint magnitude, the Owl nebula can not be seen with the naked eye. It is extremely difficult to see through binoculars, and the nebula will look like a fuzzy star through most amateur telescopes. The only way to see any detail within the gases is to use a large telescope, and only then you will just make out the two dark eyeballs of the Owl.

Messier 97 is found in Ursa Major underneath the bright star Merak, which makes up the bottom end of the pan-shape asterism.

- Suggested minimum focal length: 650mm
- Ideal focal length: 800mm to fit both, 1200mm individually

SPRING #9

DIFFICULTY: ★★★

NGC 4631
THE WHALE

The Whale Galaxy

The Whale is a barred spiral galaxy seen from its side.

The galaxy is full of details which can be seen on your final image if you spend enough time during the imaging period and do a great job with the processing.

The main challenge is to get the galaxy's yellow core, the bright blue star clusters, the red patches indicating star-forming regions, as well as the few dark dust lanes within the gases.

You will be able to capture its dwarf companion, the tiny elliptical galaxy NGC 4627, which is visible just above the Whale galaxy. If you are confident, try to capture NGC 4656/4657, the Hockey Stick galaxy (top right of left image) too!

COOL FACTS

- About the same size as the Milky Way
- Dwarf galaxy NGC 4627 located just above the Whale Galaxy
- About 28 million light-years away

DESIGNATION	NGC 4631
TYPE	Galaxy
CONSTELLATION	Canes Venatici
MAGNITUDE	9.8
SIZE	15'.5 x 2'.7

- Suggested minimum focal length: 600mm
- Ideal focal length: 1200mm

FINDING YOUR TARGET

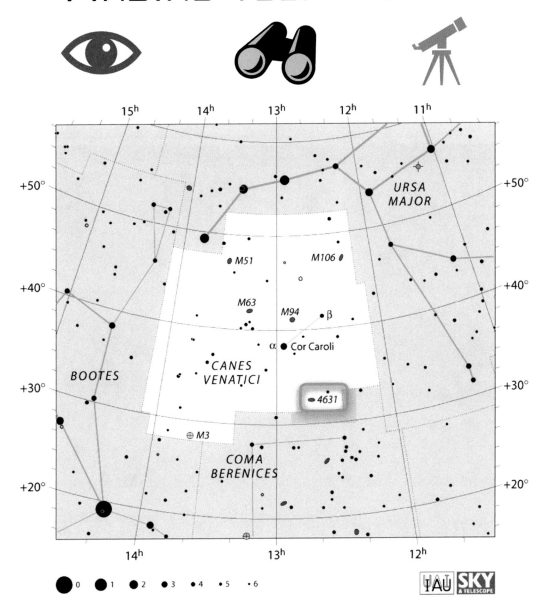

The Whale galaxy can be found in the constellation of Canes Venatici. It is too thin and dim to be seen with the naked eye and is difficult to spot with binoculars, so utilizing small telescopes to find the Whale is best due to its high surface brightness.

You may also be able to spot NGC 4627 if you are observing from an extremely dark site with no light pollution with perfect vision.

Using a large telescope will allow you to see the bright core of the galaxy and the dark spots overlaying it.

/post/whale

SPRING #10

DIFFICULTY: ★★☆

MESSIER 51
THE WHIRLPOOL

The Whirlpool Galaxy

M51 is one of the best galaxies to photograph for amateur astrophotographers who have already captured easier Messier objects (e.g. M31, M42, M45) and want to go up to the next level.

The photo on the left was taken with an 8" reflector and an unmodified DSLR camera, it is a stack of 20 x 6-minute exposure photos. The close up of M51 is from the Hubble Space Telescope. Notice how a stock DSLR camera with no filter does not show the reds seen from the image on the right. The main challenge lies in getting enough data to make out the shape of the expelling gases from their interaction. You will also be happy to see a few tiny galaxies floating in the background of your image.

COOL FACTS

- Discovered in 1773
- M51 is devouring its companion galaxy
- Three supernovas have been discovered since 1994 from the Whirlpool Galaxy

DESIGNATION	M51
TYPE	Galaxy
CONSTELLATION	Canes Venatici
MAGNITUDE	8.4
SIZE	11'.2 x 6'.9

- Suggested minimum focal length: 250mm
- Ideal focal length: 1000mm

FINDING YOUR TARGET

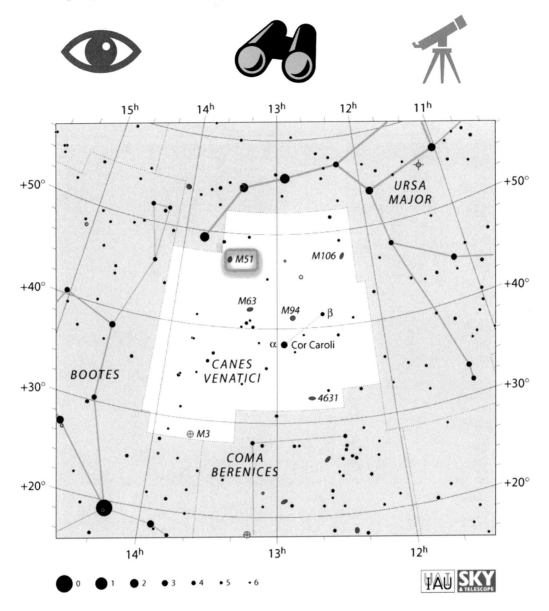

The Whirlpool Galaxy is very easy to find in the night sky. Even though it is located in the Canes Venatici constellation, you will want to use the Big Dipper asterism (in Ursa Major) to help you spot it. Simply find Alkaid, the star at the end of the Big Dipper's pan handle, and shift 3.5 degrees southwest to land on M51.

Messier 51 can be spotted with binoculars as a gray patch of light. The core of the galaxy, as well as its companion NGC 5195, can be observed through small telescopes. Using bigger instruments will let you contemplate its details within its spiral arms.

The Whirlpool galaxy is definitely one of the most impressive galaxies to capture in the sky, and it also happens to be one of the easiest targets for beginners!

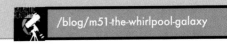
/blog/m51-the-whirlpool-galaxy

SPRING #11

DIFFICULTY: ★★★

IC 443
THE JELLYFISH

The Jellyfish Nebula

IC 443 is a remnant of a supernova that exploded between 3,000 and 30,000 years ago in our galaxy, the Milky Way.

The Jellyfish is a bit of a challenge to capture for beginners, but still deserves its place in the best 15 targets of Spring. The gases of the nebula are faint, especially the ones expanding in the larger area of the sky (see right image). Using a DSLR is not the best option for this target, but it can still yield beautiful results by spending additional time gathering data.

We recommend about 6 hours of exposure to get great details in the gases, such as the dark lanes in the expansions of gas. Make sure to keep an eye on the few bright stars in the image during processing to avoid overexposing them.

COOL FACTS

- One of the most studied supernova remnants
- 5,000 light years away from Earth
- Supernova created a Neutron Star within

DESIGNATION	IC 443
TYPE	Nebula
CONSTELLATION	Gemini
MAGNITUDE	12.0
SIZE	50'

- Suggested minimum focal length: 250mm
- Ideal focal length: 650mm

FINDING YOUR TARGET

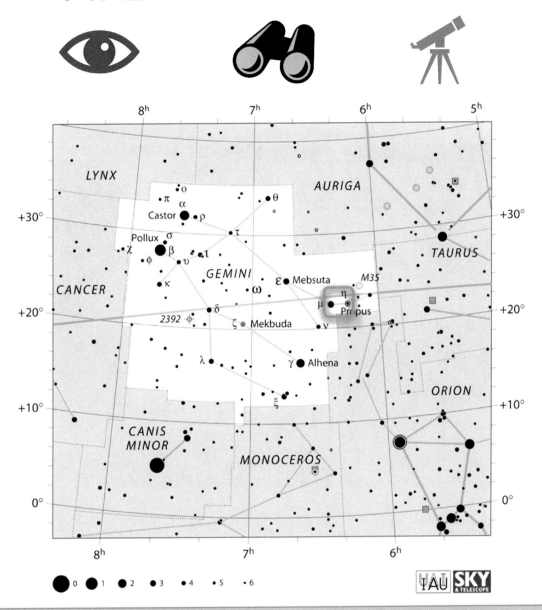

IC 443 is more of a Winter target, but does stay in the Spring skies for a little while longer than most other nebulae. It is too dim to be visible with the naked eye, but you can get a glimpse of it through binoculars if far from light polluted areas. You can also view this target with small telescopes, but the use of a filter is recommended to make the gases more obvious.

The Jellyfish nebula is located in the constellation of Gemini and is very easy to find by using the star-hopping technique.

Gemini can be divided into two sides, the left side, starting from Pollux, and the right, with Castor. The nebula is near the bottom of the Castor side. Start from this very bright star and head downward to Mebsuta, then hop to the outer star, Tejat Posterior. Now make your way to the next star (Tejat Prior), and between reaching your destination you should land on the Jellyfish. Note that the beautiful cluster of stars Messier 35 can be found near IC 443, and may be visible in the same field of view with binoculars.

SPRING #12

DIFFICULTY: ★★★

NGC 4565
THE NEEDLE

The Needle Galaxy

The Needle galaxy is the perfect example of a galaxy seen from its edge and is the largest visible one!

NGC 4565 has two companions nearby, so you can challenge yourself to have both of them visible in your final image. These neighboring bodies are dwarfed by the Needle galaxy.

The core of NGC 4565 is very luminous, so be careful during the processing time or you will ruin your image by overexposing the center. We recommend at least four hours for this target, due to its magnitude and thin size.

COOL FACTS

- Discovered in 1785
- Contains about 240 globular clusters
- Located 3 degrees away from the North Galactic Pole

DESIGNATION	NGC 4565
TYPE	Galaxy
CONSTELLATION	Coma Berenices
MAGNITUDE	9.5
SIZE	15'.90 x 1'.85

- Suggested minimum focal length: 650mm
- Ideal focal length: 1600mm

FINDING YOUR TARGET

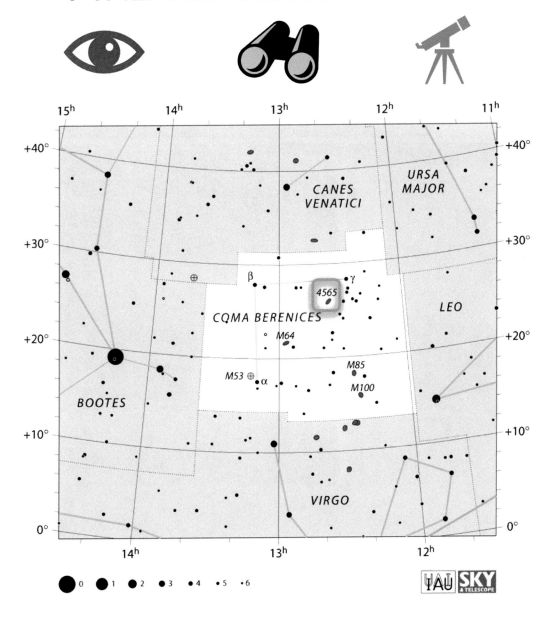

The Needle Galaxy resides in the small and faint constellation of Coma Berenices. First, locate the 5.3 magnitude star 17 Come Berenices, and you will find the galaxy about 2 degrees east. You can also easily find it by looking around the eastern edge of the naked eye Coma Star Cluster in Come Berenices.

The Needle is not visible with the naked eye. Though possible, it is a real challenge to spot with binoculars. Small telescopes will reveal a thin line of gas, similar to M82, the Cigar Galaxy.

In order to see the Needle galaxy in its full glory, a telescope of at least 16" is required. You will be able to enjoy the sight of its bright core, elongated gases, and the overlapping of dark dust lanes.

Note that this object is likely the most difficult galaxy to capture this season, but we still felt like it deserved its spot in the best targets of Spring.

SPRING #13

DIFFICULTY: ★★★

MESSIER 64
THE BLACK EYE

The Black Eye Galaxy

M64, also named the Black Eye galaxy, the Evil Eye galaxy, and the Sleeping Beauty galaxy is a spiral galaxy lying 24 million light-years away from Earth.

The reason why this galaxy got those two first evil nicknames is because of its dark band of dust that almost hides the nucleus.

Messier 64 can be a little tricky and shouldn't be rushed. While the core is bright enough to appear in any image without trouble, the gases within the galaxy are very dim and dark. On top of that, the dark dust band is hard to bring out unless a lot of time is spent gathering photons.

We recommend capturing a few galaxies in this season before confronting the Black Eye.

COOL FACTS

- Getting further at a speed of 408 km/s
- Member of the M94 group, the Cat's Eye galaxy group
- Contains about 100 billion stars

DESIGNATION	M64
TYPE	Galaxy
CONSTELLATION	Coma Berenices
MAGNITUDE	9.36
SIZE	10'.71 x 5'.13

- Suggested minimum focal length: 650mm
- Ideal focal length: 1200mm

FINDING YOUR TARGET

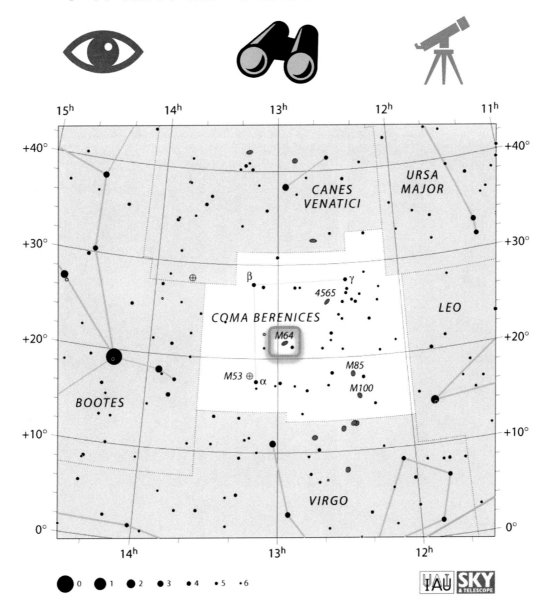

Even though M64 is pretty dim, its nucleus is very bright and easily visible in binoculars and through any telescope. The use of bigger telescopes will help you contemplate the dark dust lanes visible around the core of the Black Eye.

Although this target is very easy to see visually, finding it is a different story. Being in Coma Berenices, one of the most blended constellations in the northern hemisphere, the best way to find this target is to use the star-hopping technique. Begin on the fourth brightest star in the night sky: Arcturus, located in the constellation Bootes. From there, jump 15 degrees east to Diadem, which is the star right next to another Messier object, the globular cluster M53.

Now, picture an imaginary line from Diadem to the previous target, the Needle Galaxy NGC 4565. You will find the Sleeping Beauty in the middle of that imaginary line.

SPRING #14

DIFFICULTY: ★★☆

MESSIER 106
GALAXY IN THE HUNTING DOGS

Messier 106

The 106th object on the Messier catalog is not a very popular target because it doesn't have a nickname like most other objects. However, Messier 106 has a very unique shape which makes it one of the most beautiful galaxies out there.

Plan to spend at least 3 hours on this target, even though it would be ideal to aim for 5 hours. The key is to get enough data to bring up the shape of the arms, as well as the different colors within them.

M106 also has a companion galaxy, NGC 4217 which can be seen from its edge, so challenge yourself to capture it in the same frame!

COOL FACTS

- Two supernovas discovered since 1981
- Home to more than 400 billion stars
- Considered as being between a normal and a barred spiral galaxy

DESIGNATION	M106
TYPE	Galaxy
CONSTELLATION	Canes Venatici
MAGNITUDE	9.1
SIZE	18'.6 x 7'.2

- Suggested minimum focal length: 350mm
- Ideal focal length: 800mm

FINDING YOUR TARGET

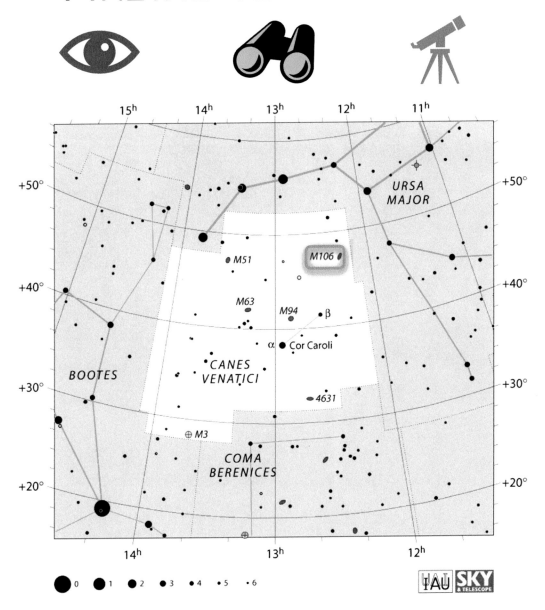

M106 can be found in the constellation of the Hunting Dogs: Canes Venatici. The easiest way to find it is to start from Ursa Major on the star forming the bottom of the Big Dipper: Phecda. Make an imaginary line southwest to Car Caroli, the brightest star in Canes Venatici, and you will find M106 about halfway through from Phecda to Car Caroli.

Like other targets mentioned in this season, M106 may be faint but has a high surface brightness making it a good target to look at. Messier 106 can be seen with binoculars and any telescope. Binoculars will show a faint patch of light, whereas telescopes will reveal a little bit of structure and even some details in the spiral arms depending on the size of the instrument. We recommend spending some extra time with a Hydrogen Alpha filter for this target, as it is rich in glowing-red star forming regions.

/blog/_m106

SPRING #15

DIFFICULTY: ★★★

NGC 2359
THOR'S HELMET

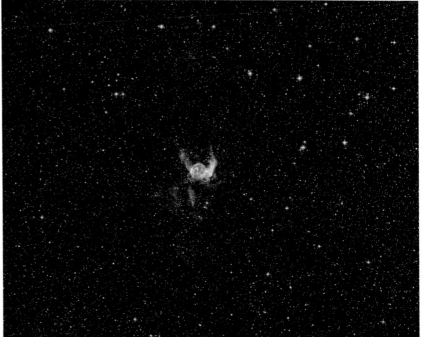

Thor's Helmet

Thor's Helmet is a cloud of interstellar gas that resembles Thor's Helmet. Although it is a faint target, the colors in the gases really pop when taking long exposures, even with a DSLR camera that is not modified.

This beautiful deep sky object gets its glow from WR7, a massive Wolf-Rayet star that will soon turn into a supernova.

We recommend as many hours as your patience can allow to capture this target. While a total of 4 hours can yield fair results, additional time will give you all the faint gases surrounding NGC 2359.

COOL FACTS

- About 30 light-years across
- Wolf-Rayet star gives NGC 2359 its glow
- Similar to the Bubble Nebula, but more complex

DESIGNATION	NGC 2359
TYPE	Nebula
CONSTELLATION	Canis Major
MAGNITUDE	11.45
SIZE	8' x 8'

- Suggested minimum focal length: 500mm
- Ideal focal length: 1000mm

FINDING YOUR TARGET

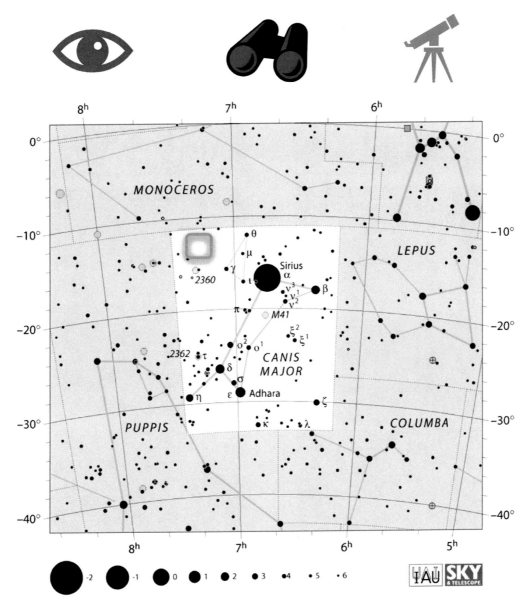

NGC 2359 can be found in the constellation of the Great Dog, Canis Major. It is fairly easy to find, because it is very close to the brightest star in the night sky, Sirius. From there, simply travel about 8 degrees northeast in order to find the nebula.

This object is also better captured in Winter, but, like the Rosette and Jellyfish nebula, stays available in the sky for longer than other nebulae.

Thor's Helmet is a faint target that is impossible to spot with the naked eye or binoculars. A small beginner telescope can spot its nebulosity from an extremely dark site, but do not expect to view something impressive. A bigger instrument will reveal some of the shapes in the gases. If you would like to see more of the nebula, you would need to attach a filter to a high power telescope, and only then will you be able to distinguish the iconic shape of the helmet.

/post/thorshelmet-ngc2359

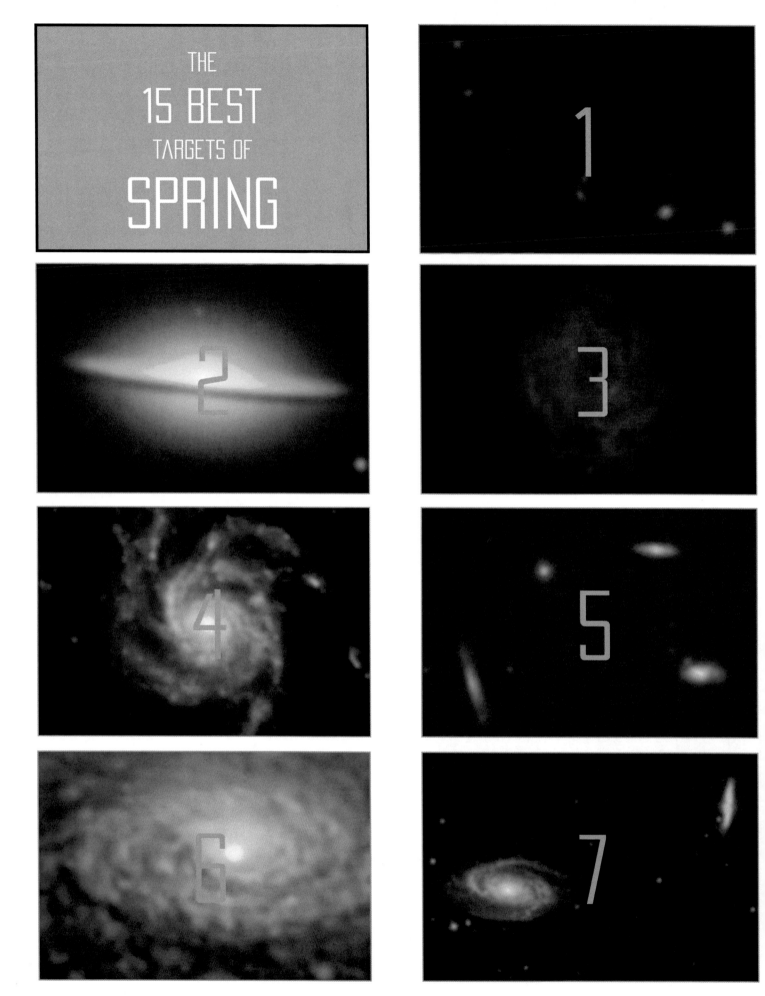

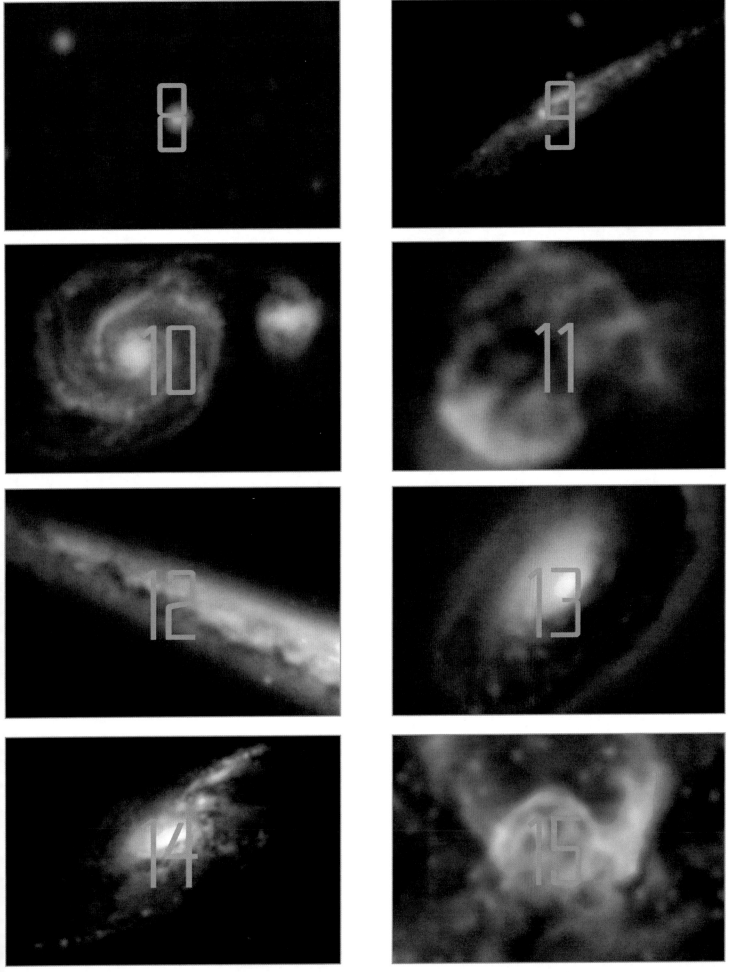

THE 15 BEST TARGETS OF... SUM

MER

GOT ENOUGH GALAXIES DURING SPRING? GREAT, BECAUSE SUMMER IS RIDDLED WITH THEM!

THIS SEASON HOLDS SOME OF THE MOST ICONIC DEEP SKY OBJECTS, SUCH AS THE PILLARS OF CREATION, THE VEIL NEBULA, THE GREAT CLUSTER IN HERCULES, AND MORE!

THE GOOD THING IS, ALL THE TARGETS LISTED BELOW CAN EASILY BE PHOTOGRAPHED BY BEGINNERS.

MESSIER 8
THE LAGOON

DIFFICULTY: ★☆☆

The Lagoon Nebula

We start the Summer season with the large M8. The Lagoon Nebula is the 8th deep sky object in the Messier Catalog, and is a great target because of its size, magnitude, and color.

If imaging from a dark location, you will not have to spend much time on this target. The right photo was taken with an unmodified DSLR camera and only a total of 2 hours. A modded DSLR camera may show more gas but make the entire target red. Without filters or modification, you can differentiate the blues from the reds, but will capture less outer gases. The choice is yours!

Can you spot the open cluster NGC 6530? It is in front of the left side of the Lagoon Nebula. The cluster was formed from the gases of M8.

COOL FACTS

- Open Star Cluster NGC 6530 in front of the nebula
- Discovered before 1654
- Bright center contains the Hourglass nebula

DESIGNATION	M8
TYPE	Nebula
CONSTELLATION	Sagittarius
MAGNITUDE	6.0
SIZE	90' x 40'

- Suggested minimum focal length: 135mm
- Ideal focal length: 650mm

FINDING YOUR TARGET

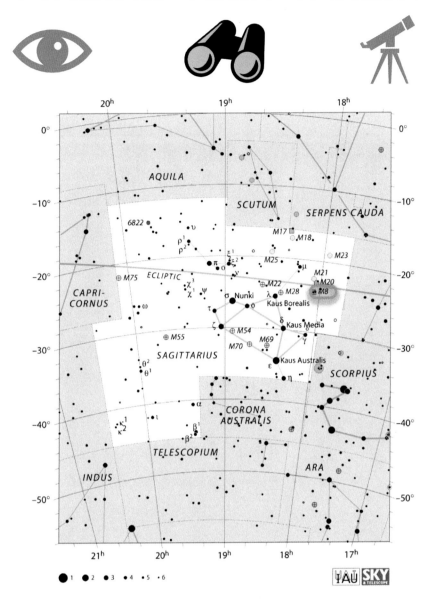

The Lagoon nebula is located in the Sagittarius constellation. You can find it in the busy Milky Way band, above the top right of Sagittarius' teapot asterism.

Make sure to plan ahead before going out for imaging this target, as the nebula does not rise very high in the sky in the Northern hemisphere.

You can easily see both the Lagoon nebula and its star cluster with binoculars from a dark zone. It is also visible to the naked eye as a gray patch, but a little hard to distinguish since it is inside the Milky Way band. A telescope will reveal visible details like darker shades of gray within the gases.

If you own a monochrome camera, this is a great narrowband target, especially when the channels are combined as "SHO" (Hubble Palette).

/post/m8-the-lagoon-nebula

DIFFICULTY: ★★☆

NGC 6888
THE CRESCENT

The Crescent Nebula

NGC 6888 is a colorful target that is suitable to image with telescopes of any focal length.

The Crescent Nebula is lit up by its central star, a massive Wolf-Rayet star, that blows stellar winds all around. The star is nearing the end of its life and should produce a massive supernova explosion when it dies.

Photographing and processing this nebula is fun because of the gases and hints of nebulosity. If you do not have narrowband filters, you should image from a site free of light pollution early in the summer.

Although too faint and difficult to be in this list, try to get its neighbor, the Soap Bubble Nebula (left image, bottom middle), in the same frame!

COOL FACTS

- Discovered in 1792
- Also called the Euro Sign Nebula
- Soap Bubble Nebula nearby, discovered in 2007 by an amateur astrophotographer

DESIGNATION	NGC 6888
TYPE	Nebula
CONSTELLATION	Cygnus
MAGNITUDE	7.4
SIZE	18' x 12'

- Suggested minimum focal length: 85mm
- Ideal focal length: 1200mm

FINDING YOUR TARGET

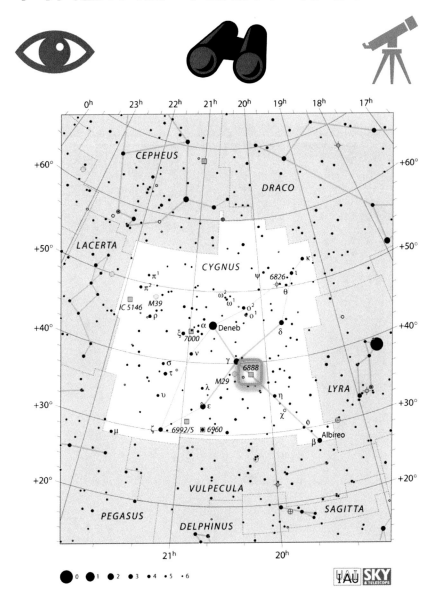

The crescent nebula is not bright enough to be seen with the unaided eye. You could have a glimpse of it with binoculars, only if a filter is added, but you wouldn't even be sure that you are looking at the right target. A 10"+ telescope with a wide field eyepiece is recommended to see the gases forming the Crescent Nebula.

If you are not able to see it directly, try looking with averted vision, or try gently tapping the telescope, as the gases will be more visible if your vision is not completely still.

The Crescent nebula can be found in the constellation Cygnus, between the middle star, Sadr, and the star at the base of the swan's neck. Without the use of filters, be sure to image or look for this target from a very dark site only. The Crescent nebula is surrounded by hydrogen alpha gas, and makes for a fantastic narrowband target even with small telescopes.

/post/ngc-6888-the-crescent-nebula

DIFFICULTY: ★☆☆

MESSIER 20
THE TRIFID

The Trifid Nebula

Messier 20 is great for any instrument. A close-up image will show a lot of details, especially in Barnard 85, the dark nebula. A wide shot will allow you to capture both M20 and M8 in the same frame. Both are about the same magnitude and make for an impressive photo together.

Just like M8, M20 doesn't get very high from the horizon. This can be challenging when photographing it because you want to make sure that there are no big cities in the direction of the nebula, or it will create a light pollution dome that you will have to shoot through.

You also don't need to spend a lot of time on this target in order to get great results. The image on the right shows a similar exposure time to M8, about 2 hours total. Check out M21 to the right!

COOL FACTS

- Dark nebula in front appears to separate the two blobs
- Photographed by Hubble in 1997
- Brightest star is a triple star system

DESIGNATION	M20
TYPE	Nebula
CONSTELLATION	Sagittarius
MAGNITUDE	6.3
SIZE	28'

- Suggested minimum focal length: 200mm
- Ideal focal length: 800mm

FINDING YOUR TARGET

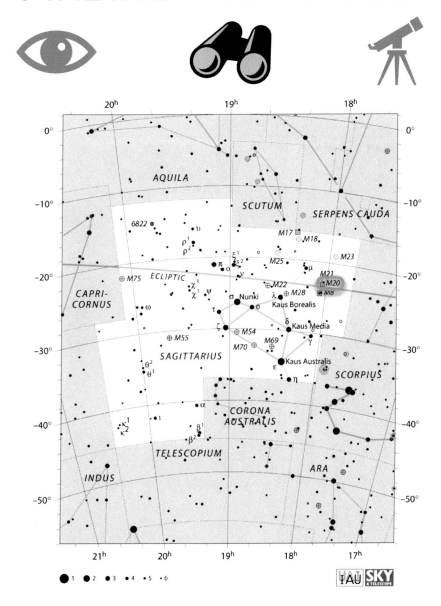

M20 is located just above the Lagoon nebula, in the Milky Way band. To find it, you can simply follow the same steps as M8, but this time head just a bit higher.

Messier 20 is smaller than its neighbor and will be easy to spot with binoculars, but seeing it with the naked eye is more of a guessing game. It will look more like a fuzzy star than a gray patch.

Once again, do not wait until the last minute to capture this nebula, as it does not rise very high.

You may also see Messier 21 through binoculars or a telescope. M21 is a star cluster located very close to the Trifid!

Surprisingly, the blue blob in M20 barely shows when shooting it in narrowband, so it is easier to reveal the entire object through simple broadband astrophotography.

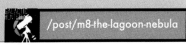
/post/m8-the-lagoon-nebula

MESSIER 27
THE DUMBBELL

DIFFICULTY: ★☆☆

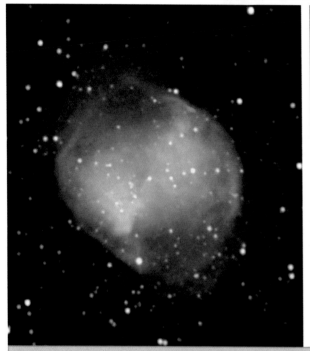
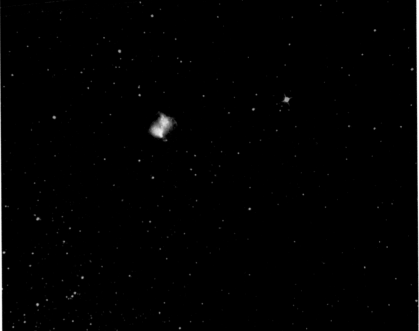

The Dumbbell Nebula

M27 was the first planetary object discovered by Charles Messier in 1764.

This target does not require many hours to get great results. The photo on the right was taken with only 1 hour and a half worth of 6 minute exposures, and you can easily distinguish the shape of a dumbbell.

The biggest challenge is to bring out the red "X" striking in the center of the nebula. This is not easy task with a stock DSLR unless you spend a lot of time on it, but it is achievable. The key is to image from a zone that is dark enough, and make sure to bring out the red gases during processing. As always with emission nebulae, a duo band or individual narrowband filters will help!

COOL FACTS

- Inner gas has the shape of a dumbbell
- Very similar to what our sun will become when it dies
- Second brightest planetary nebula

DESIGNATION	M27
TYPE	Nebula
CONSTELLATION	Vulpecula
MAGNITUDE	7.5
SIZE	8' x 5'.6

- Suggested minimum focal length: 350mm
- Ideal focal length: 1200mm

FINDING YOUR TARGET

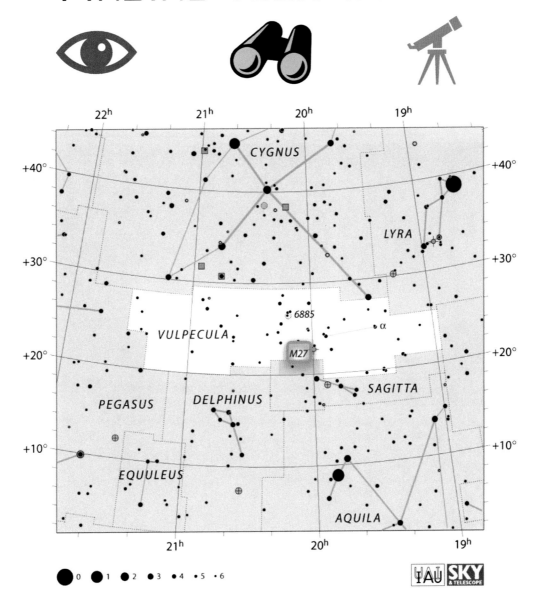

Messier 27 is the second brightest planetary nebula, and large enough to be visible with binoculars and small telescopes. Due to its high surface brightness, a telescope will yield more details in the gases of the nebula, and you may even recognize the shape of a dumbbell!

The Dumbbell nebula can be found inside the famous Summer Triangle (composed of the Altair, Deneb, and Vega), in the constellation of Vulpecula. Star hop from Altair and head down towards Deneb in a straight line. You will land on M27 about one-third of the way there.

Also note that an open cluster, NGC 6830 containing just about 30 stars lies just a couple degrees west of the nebula.

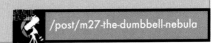
/post/m27-the-dumbbell-nebula

MESSIER 17
THE OMEGA

DIFFICULTY: ★★☆

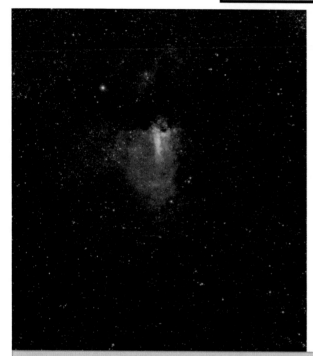
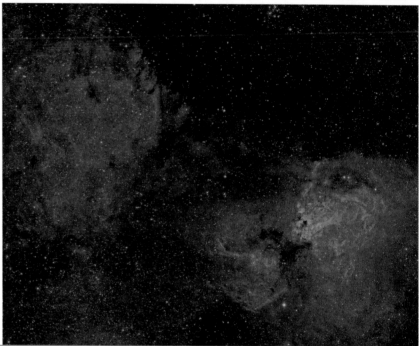

The Omega Nebula

M17 is a fairly bright diffuse emission nebula, and is also commonly named the Swan nebula.

M17 is very similar to the famous Orion nebula (M42), but it is seen from its edge, while M42 is seen face-on. Messier 17 is also about three to four times farther than the Orion nebula.

This target is easy to photograph, but gets a 2-star difficulty rating because of the surrounding gases. Getting the bright part of M17 is an easy feat, but if you would like to get some of the faint outer gases, make sure to spend some more time on gathering enough data. You will also want to keep those fainter gases in mind during the entire processing, so you don't accidentally make your image too dark and lose the color of the gases all around.

COOL FACTS

▸ Discovered in 1745
▸ One of the brightest star-forming nebulae
▸ Home to 800 stars

DESIGNATION	M17
TYPE	Nebula
CONSTELLATION	Sagittarius
MAGNITUDE	6.0
SIZE	11'

▸ Suggested minimum focal length: 50mm
▸ Ideal focal length: 400mm

FINDING YOUR TARGET

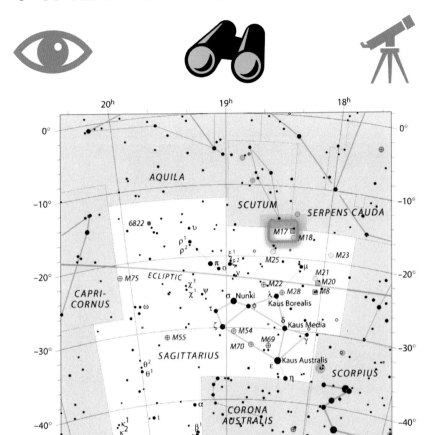

Messier 17 is best seen through wide telescopes and binoculars. Only people with very good vision will be able to spot it with the naked eye under an extremely dark skies without light pollution.

The Omega nebula is located in the constellation of Sagittarius, near the Eagle Nebula. You can find it easily by spotting the bottom star of the constellation Scutum, or also by using the teapot stars in Sagittarius.

If using binoculars when trying to find it, make sure not to mistake the Omega Nebula with the Eagle Nebula, the Trifid or Lagoon nebulae, because they are all close to each other.

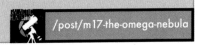

DIFFICULTY: ★★★

MESSIER 57
THE RING

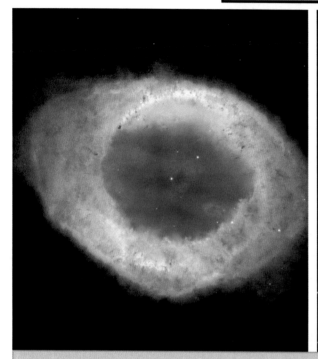

The Ring Nebula

M57 is a tricky target to capture!

The two main goals when photographing the Ring nebula are to:

1. Get the white dwarf star in the center to be visible, which is not difficult.
2. Have both the inner and outer gases in the final image.

The outer gases are the faint bits of nebulosity being expelled from the ring. While not too hard to see once all frames are stacked, it is difficult to keep those gases without blowing out the core of the nebula.

To get proper results, we recommend more than 4 hours on this target.

COOL FACTS
- One of the smallest Messier objects
- Growing at a rate of 43,000+ mph
- Will faint away in 10,000 years

DESIGNATION	M57
TYPE	Nebula
CONSTELLATION	Lyra
MAGNITUDE	8.8
SIZE	230" x 230"

- Suggested minimum focal length: 800mm
- Ideal focal length: 1600mm

FINDING YOUR TARGET

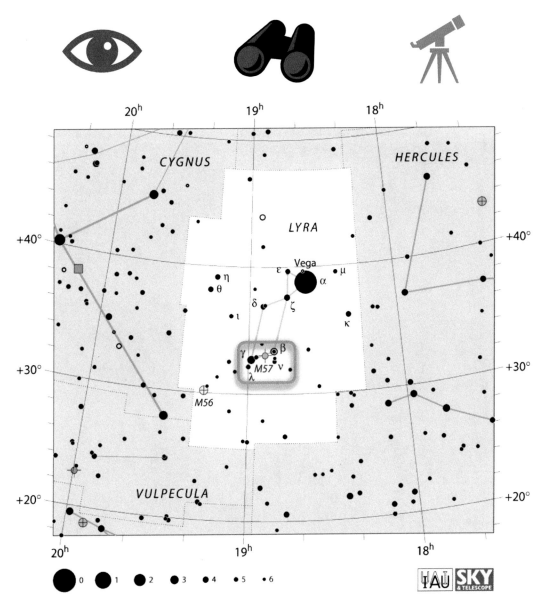

The Ring nebula lies just 2,283 light-years away from Earth, just south of the bright star Vega, in the constellation of Lyra.

Messier 57 is very easy to find, because it is located almost exactly in between the two bottom stars of Lyra, Sheliak and Sulafat. Start from either one of those two and make your way to the other in a straight line. You will find your target a little closer to Sheliak than exactly in the center.

Because of its tiny size, M57 cannot be seen with the naked eye or even binoculars. You may be surprised by the ring's brightness when looking at it through a telescope. It may be tiny but it is a wonderful object to view!

/post/m57-the-ring-nebula

DIFFICULTY: ★★☆

MESSIER 24
SAGITTARIUS STAR CLOUD

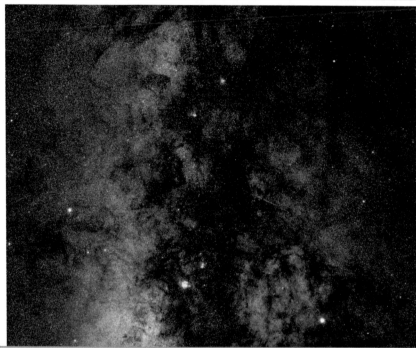

The Sagittarius Star Cloud

M24 is a large and bright star cloud, spanning 600 light years wide in the sky.

This is a good target for DSLR cameras on a star tracker, or attached to a wide telescope. What makes this star cloud impressive, aside from the tens of thousands of stars visible, are the several interstellar dust lanes that divide some of these packs of stars.

The main challenges you will face are finding the perfect area of the sky to point your camera at, and processing your results.

Make sure to go easy when editing your image, because it doesn't take much to turn this beautiful blanket of stars into an over processed mess.

COOL FACTS

- About 9 times the size of the Moon
- Cluster NGC 6603 in the brightest part of the cloud
- Not considered a Deep Sky Object

DESIGNATION	M24
TYPE	Star Cloud
CONSTELLATION	Sagittarius
MAGNITUDE	4.6
SIZE	90'

- Suggested minimum focal length: 24mm
- Ideal focal length: 85mm

FINDING YOUR TARGET

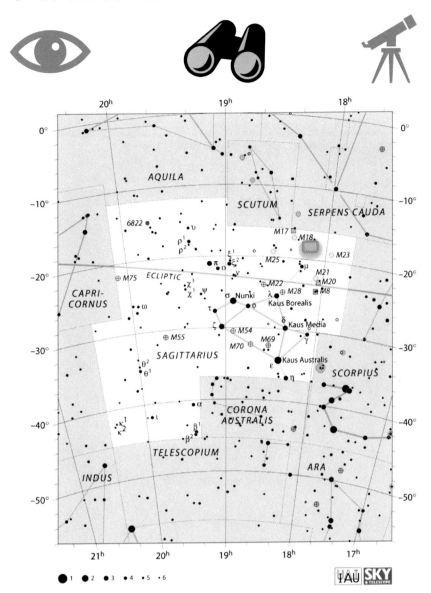

Excluding clusters, M24 is the densest concentration of stars in the Messier catalog. Through binoculars, 1000+ stars can be seen in just one spot. The star cloud is also a good visual target for low-power telescopes, and is one of the few most impressive targets through binoculars. You may also see M24 with the naked eye without much trouble from a zone far from light pollution.

The Sagittarius Star Cloud can be seen anytime when the Milky Way is visible, so the best time to view it is during the Summer months. It can be found in the Sagittarius constellation, near M17, the Omega Nebula, and the open cluster M18. It is located about 4 degrees north of the bright blue star above the Teapot asterism.

DIFFICULTY: ★★☆

MESSIER 16
THE EAGLE

The Eagle Nebula

M16, the Eagle Nebula, is one of the most well-known nebulae in the night sky since the Hubble Space Telescope photographed its Pillars of Creation, in 1995 (left).

The cropped image on the right shows the nebula photographed with a stock DSLR through an 8" telescope. You can see the Pillars of Creation in the center, the beak of the Eagle above, and the faint, spanned wings on its sides.

You should spend a minimum of 4 hours on this target to get a similar result with a DSLR camera, but adding more time will help capture more of the outer gases forming the wings of the Eagle.

COOL FACTS

- Discovered in 1745 and thought to be a star cluster
- Made famous by the HST in 1995
- Tallest pillar is 4 light-years high

DESIGNATION	M16
TYPE	Nebula
CONSTELLATION	Serpens
MAGNITUDE	6.0
SIZE	30'

- Suggested minimum focal length: 250mm
- Ideal focal length: 650mm

FINDING YOUR TARGET

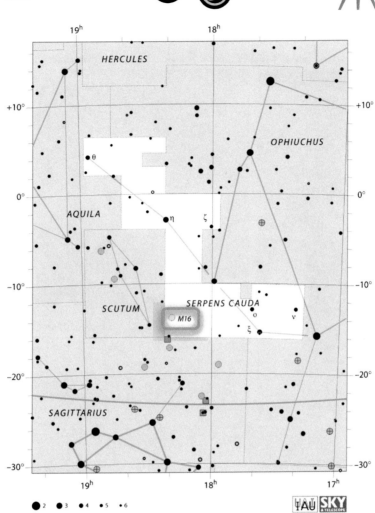

M16 can be found in the tail of the Serpens constellation near Scutum. It lies about 2.5 degrees west of the bright star Gamma Scuti.

The Eagle nebula is close enough to M17, the Omega nebula, that they can both be seen in the same field of view when using binoculars. M16 is not very impressive through binoculars. The cluster of stars will easily be seen, but the gases forming the nebulosity will be far from obvious.

The Eagle nebula as a whole is best seen through small telescopes. The Pillars of Creation, being much smaller, are best seen with along focal length telescope and might benefit from the use of a filter.

 /post/m16-the-eagle-nebula-the-pillars-of-creation

DIFFICULTY: ★★★

XSS J16271-2423
RHO OPHIUCHI

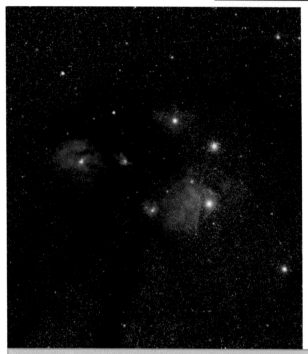

Rho Ophiuchi

At a distance of about 400 light years, the binary star system Rho Ophiuchi is the closest stellar nursery to Earth. It surrounds the huge orange star Antares.

Because of its size, you will not need to capture this target with a telescope. The challenge is to get as much nebulosity as possible without the bright stars being blown out.

We recommend 3 minute exposures with a lens of 50mm-135mm for this target. Be sure to photograph it when it gets high enough in the sky, which is not for long. If not, your image will be affected by atmospheric turbulence and/or light pollution from the light dome of nearby cities.

COOL FACTS

- Consists of 2 major regions of gas and dust
- Temperatures within the clouds range from 13K to 22K
- Good to photograph with Saturn if present

DESIGNATION	XSS J16271-2423
TYPE	Star System
CONSTELLATION	Ophiuchus
MAGNITUDE	4.6
SIZE	4.5° × 6.5°

- Suggested minimum focal length: 50mm
- Ideal focal length: 135mm

FINDING YOUR TARGET

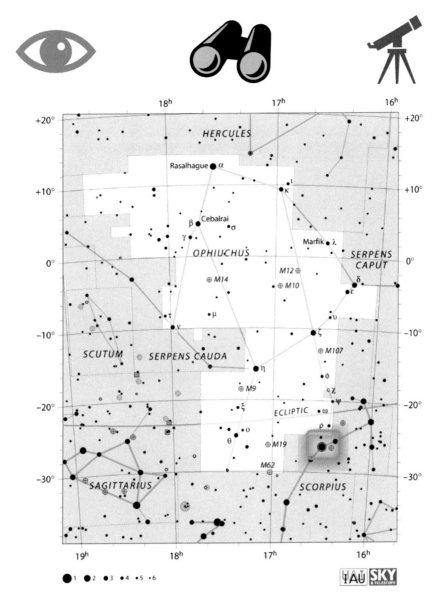

Rho Ophiuchi covers an angular area of 4.5° × 6.5° in the night sky, which is huge! For comparison, the full moon is about 0.5°. This is why you will not be able to see this target using a telescope.

This is the easiest target to locate out of all targets listed in this guide. To find it, look towards Antares, the brightest star in the constellation Scorpius. It is also the brightest and most orange star in the entire Summer sky.

Pointing your camera at Antares assures that you will capture the clouds of Rho Ophiuchi, as Antares lies within the complex. All you would have to do next is to center the field of view to your liking. If using a 50mm lens, you can capture some of the Milky Way band on one side of your image while still keeping Rho Ophiuchi visible.

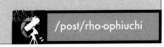
/post/rho-ophiuchi

NGC 7000 & IC 5070
NORTH AMERICA & PELICAN

DIFFICULTY: ★★★

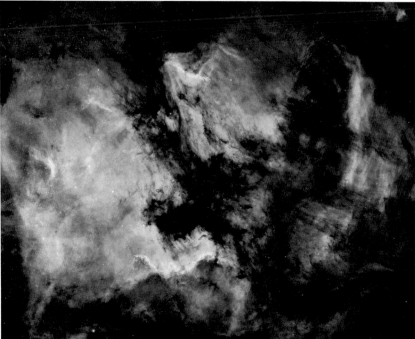

North America & Pelican Nebulae

These two huge emission nebulae are separated by a wall of dark nebulosity, and are great for any type of instrument!

The North America nebula (left) is more than four times the size of the moon, and looks like the North American continent. The Pelican Nebula is fainter and smaller, but appears very well in photographs. Both can be captured in the same field of view if shooting wide-field, as seen on the right image.

If you are not using filters, you will get something mostly red like the picture on the left. Using narrowband filters or a single duo-band filter will give you a wider range of colors to work with.

COOL FACTS

- Discovered in 1786
- Bright stars form the "Little Orion" asterism
- Cygnus Wall divides both targets with dark nebulosity

DESIGNATION	NGC 7000 IC 5070
TYPE	Nebulae
CONSTELLATION	Cygnus
MAGNITUDE	NGC 7000 - 4 IC 5070 - 8
SIZE	NGC 7000 - 120' x 100' IC 5070 - 60' x 50'

FINDING YOUR TARGET

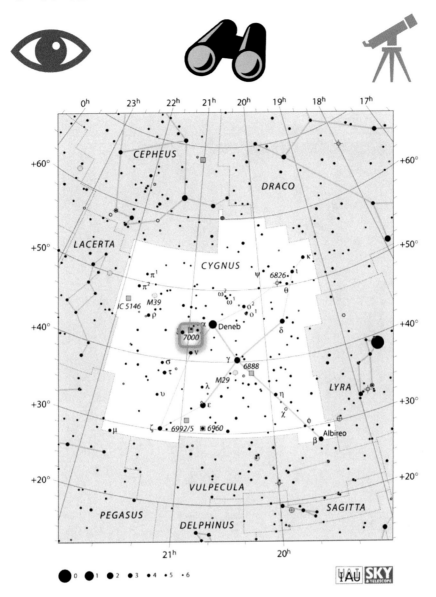

Both nebulae are located in the constellation of Cygnus, the swan. The star at the tail end of Cygnus is its brightest star, Deneb, and moving westward you will find your targets.

Despite being very large, you will not be able to spot them with the naked eye, as their gases are spread out and quite dim. However, a pair of binoculars or small telescope will reveal a patch of gray light. For better results, the use of an UHC filter is recommended when looking at the target through a telescope, but you will not be able to see any shapes or details until taking a long exposure shot with a camera.

This is a fantastic target for the Hubble Palette combination (SHO) as it will show so much more detail all over the area than if doing a true-color image.

- Suggested minimum focal length: 135mm
- Ideal focal length: 350mm

DIFFICULTY: ★★☆

MESSIER 11
THE WILD DUCK

The Wild Duck Cluster

M11, the Wild Duck Cluster, got its name because the brightest stars in the cluster seem to form a triangle resembling a flock of ducks flying.

It is one of the most compact and star-rich clusters out there, with about 2,900 stars.

The image on the right is what you can expect when photographing the Wild Duck with a telescope that has a focal length of 800mm. The exposure time was just 2 hours. Notice that you can see dark lanes in several spots around the cluster.

Make sure your tracking and guiding are perfect for this target, as with any compact cluster, or you will end up with a blurry mess.

COOL FACTS

- One of the most populated clusters known
- About 250 million years old
- Discovered in 1733

DESIGNATION	M11
TYPE	Cluster
CONSTELLATION	Scutum
MAGNITUDE	6.3
SIZE	14'

- Suggested minimum focal length: 650mm
- Ideal focal length: 1000mm

FINDING YOUR TARGET

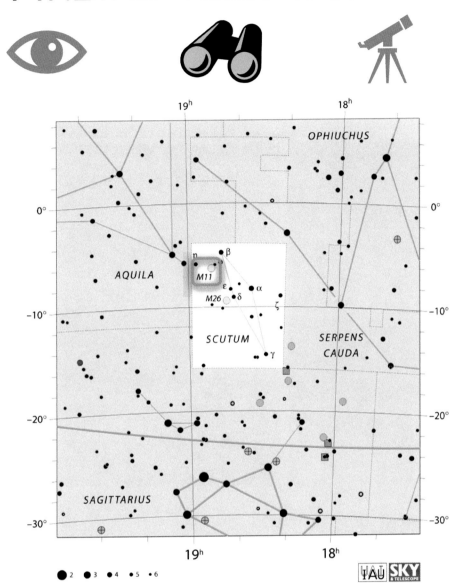

At a distance of 6,200 light-years, Messier 11 is the farthest open cluster in the Messier catalog that can be seen with the naked eye. The Wild Duck cluster is also a good target with binoculars and will appear as a bright but fuzzy diamond. You might be able to distinguish the triangle, or V-shape of the cluster through a telescope.

M11 is located in the constellation of Scutum, not too far from another Messier cluster, M26. If you are having trouble finding Scutum, you can also spot the Wild Duck Cluster above the Teapot asterism from the Sagittarius constellation.

/post/m11-the-wild-duck-cluster

DIFFICULTY: ★★★

NGC 7380
THE WIZARD

The Wizard Nebula

Located approximately 8,000 light-years away, NGC 7380 is an open cluster with nebulosity forming the shape of a medieval wizard.

The Wizard nebula is best photographed with narrowband filters, as it is filled with HA and OIII. It is still a great target for stock DSLR camera owners, but expect to spend long hours during your imaging session to get fair results.

For the best result using a DSLR/Mirrorless camera, spend a few hours imaging this target from a dark site, then spend some extra hours gathering more data but this time using a clip-on Hydrogen-Alpha filter. You can then combine it all and process it to get impressive results.

COOL FACTS

- Discovered by William Hershel's sister, Caroline
- Will only be visible for a few million years
- Spans about 100 light-years

DESIGNATION	NGC 7380
TYPE	Nebula
CONSTELLATION	Cepheus
MAGNITUDE	7.2
SIZE	25'

- Suggested minimum focal length: 135mm
- Ideal focal length: 800mm

FINDING YOUR TARGET

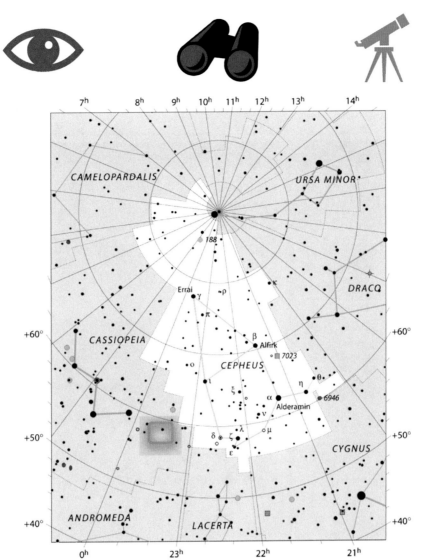

The Wizard nebula is located in the Cepheus constellation, on the right side of Cassiopeia's "W" shape.

Finding it is difficult, but the best way to do so is to imagine a straight line between Cepheus' bottom left star and Cassiopeia's right star. The nebula will be on that line but closer to Cassiopeia than Cepheus.

Do not expect to spot the Wizard with your naked eye or binoculars, because it is not possible even in very dark skies. The only way to see the nebula is through a telescope. We recommend observing far from light polluted areas and consider the use of an O-III filter.

/post/ngc-7380-the-wizard-nebula

DIFFICULTY: ★★☆

MESSIER 13
THE GREAT GLOBULAR CLUSTER IN HERCULES

The Great Globular Cluster in Hercules

Messier 13 is one of the best globular cluster to observe and photograph from the Northern hemisphere. It is bright, large, and stays high in the sky for a long period of time.

Photographing this cluster is easy and does not require a lot of integration time. For the best results, keep your exposure times short to avoid blowing up the bright stars, and always check that your guiding is perfect.

The photo on the right was taken with a DSLR camera and an 800mm focal length telescope, for two hours total of exposure time. The image on the left is from NASA/ESA.

COOL FACTS
- Charles Messier thought it was a nebula
- Cluster is half a million solar masses
- About 11.65 billion years old

DESIGNATION	M13
TYPE	Cluster
CONSTELLATION	Hercules
MAGNITUDE	5.8
SIZE	20'

- Suggested minimum focal length: 300mm
- Ideal focal length: 1000mm

FINDING YOUR TARGET

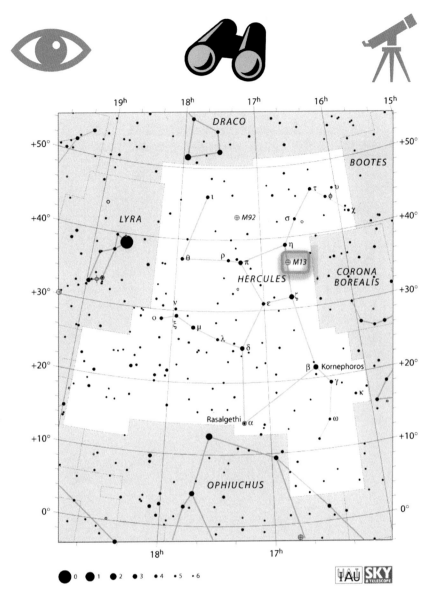

The Great Globular Cluster in Hercules can be seen with telescopes, binoculars, and the naked eye! The latter must be seen from a dark zone with no light pollution. Through a telescope, the cluster really looks impressive and many stars can be resolved.

You will find the cluster in the constellation of Hercules and have no issue spotting it! Once you find it, take a moment to think about how Charles Messier found the same cluster many years ago through his own telescope. With his instrument, M13 (and most other objects at the time) looked like a nebula!

This is his entry for his famous catalog after having discovered the cluster: *"In the night of June 1 to 2, 1764, I have discovered a nebula in the girdle of Hercules, of which I am sure it doesn't contain any star; having examined it with a Newtonian telescope of four feet and a half, which magnified 60 times, it is round, beautiful & brilliant, the center brighter than the borders"*.

/post/m13-the-great-globular-cluster-in-hercules

DIFFICULTY: ★★☆

NGC 6960
THE VEIL

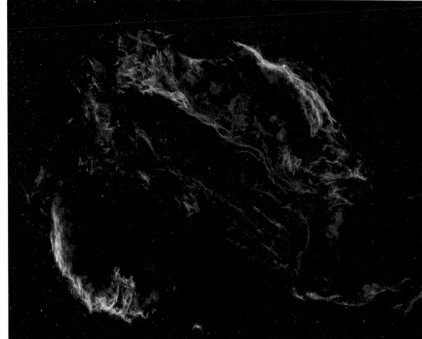

The Veil Nebula

NGC 6960, or the Veil Nebula, is also often called the Witch's Broom nebula or Bridal Veil nebula. It is a beautiful diffuse nebula divided into three parts: The Eastern Veil (right image, bottom left side), the Western Veil (left image), and Pickering's Triangle (right image, top center).

Photographing each part of the Veil is simple, though Pickering's Triangle is fainter than the two others. The photo of the Western Veil (left) was taken through an 8" telescope, with about 3 hours total exposure time without any filter. The

The image on the right was taken with a small refractor telescope and two narrowband filters, Hydrogen Alpha and Oxygen III.

COOL FACTS

- Discovered in 1784
- Cloud of heated and ionized gas
- Large supernova remnant that constitutes the Cygnus Loop

DESIGNATION	Eastern: NGC 6992 Western: NGC 6960 Triangle: NGC 6979
TYPE	Nebula
CONSTELLATION	Cygnus
MAGNITUDE	7.0
SIZE	3°

FINDING YOUR TARGET

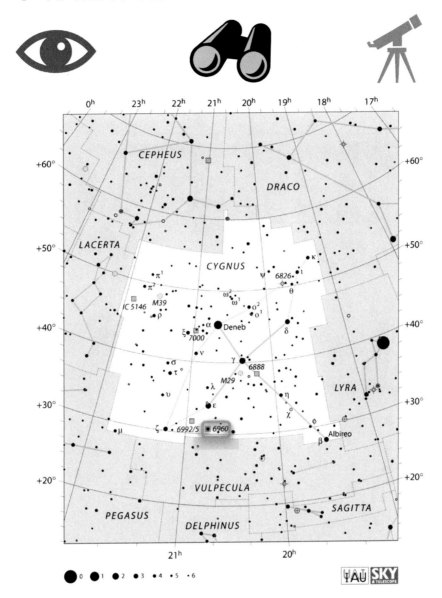

The Veil nebula is located in Cygnus, just a few degrees south of the star Gienah, in the right wing of the swan.

The Western Veil is the easiest to spot because of its bright star (52 Cygni) that can quickly be found with the naked eye. This target can only be seen through binoculars or a telescope and will be difficult to spot without a OIII or UHC filter. A 6" to 10" will reveal a blurry, elongated haze, while a larger aperture telescope will allow you to resolve the gas filaments of the nebula.

The Western and Eastern veils are worth staring at through an instrument, but Pickering's triangle is much too faint and will be a challenge to observe.

- Suggested minimum focal length: 100mm
- Ideal focal length: 350mm

DIFFICULTY: ★★★

MESSIER 75
GLOBULAR CLUSTER IN SAGITTARIUS

Messier 75

Another great globular cluster to photograph during the summer season is M75, which is much smaller than M15 (6' vs. 18') and dimmer (6.2 vs 9.18, remember: the lower the number, the brighter the object), but still impressive!

Once again, make sure you do not take long exposures to avoid blowing out the bright stars, and continuously check that your tracking and guiding are perfect to avoid any star trails. Also plan ahead and confirm that there will be no wind during the night you are imaging. Gusts of wind can distort each photo causing this magnificent ball of stars to become a blurry white mess!

It is often advised to completely turn off the auto-guiding so that you do not chase the Seeing.

COOL FACTS

- Discovered in 1780
- Does not have a cool nickname
- One of the most densely concentrated Globular Clusters known

DESIGNATION	M75
TYPE	Cluster
CONSTELLATION	Sagittarius
MAGNITUDE	9.18
SIZE	6'.8

- Suggested minimum focal length: 350mm
- Ideal focal length: 1000mm

FINDING YOUR TARGET

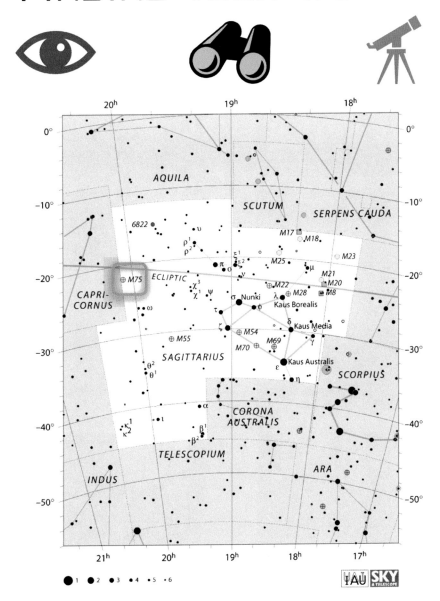

Because of its size and distance of 47,600 light years, Messier 75 is not visible to the naked eye. It is the second most distant Messier globular cluster after M54. Through binoculars, the cluster appears as a faint star and is barely visible. The best way to see this target is to use a long focal length telescope, so you will be able to distinguish the stars within the cluster.

M75 is located in the western area of the Sagittarius constellation, and has an apparent size of 6.8 arc minutes. Finding it without a GoTo telescope is a challenge, as it sits in between Sagittarius and Capricornus, far from any bright star.

Because star hoping is not really a good option here, the best way to land on the cluster is to follow the Ecliptic line from Sagittarius to Capricornus.

THE 15 BEST TARGETS OF SUMMER

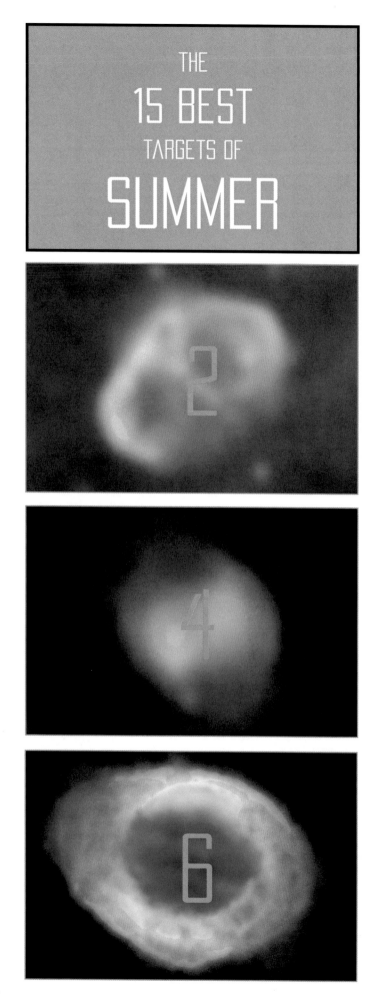
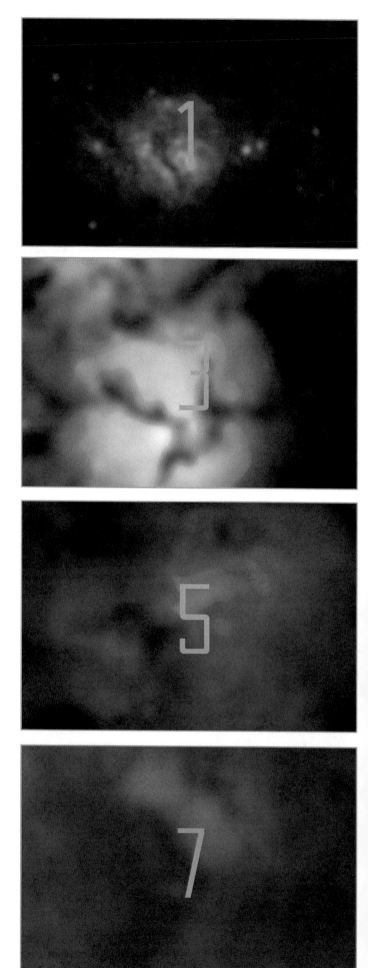

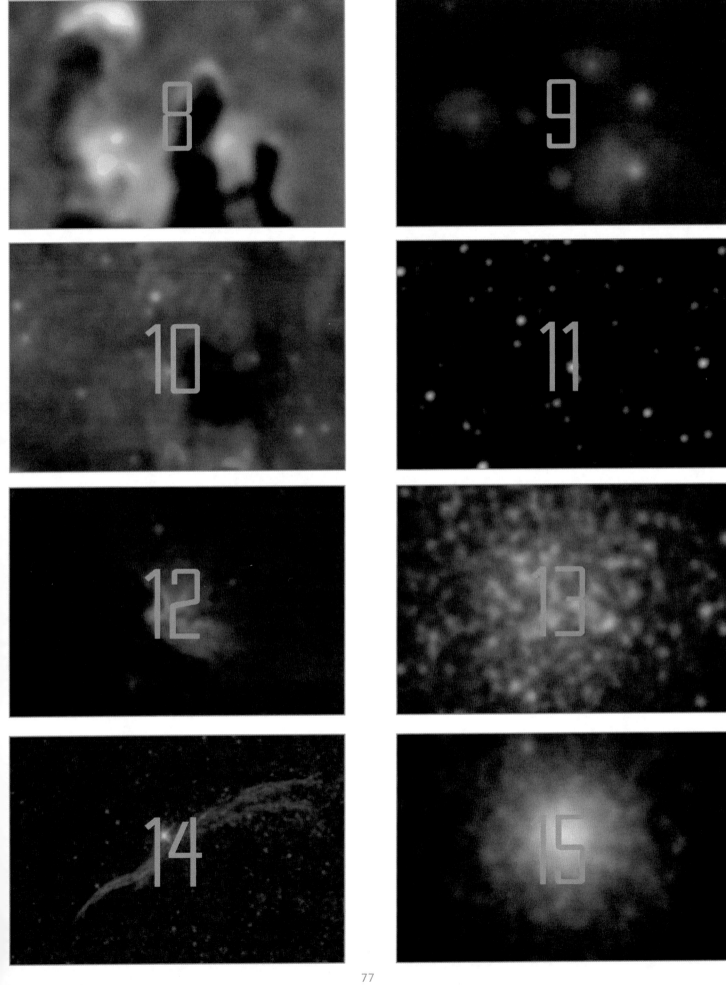

THE 15 BEST TARGETS OF...

FA

FALL

FALL HAS A VARIETY OF DEEP SKY OBJECTS THAT ARE WONDERFUL FOR PHOTOGRAPHY, INCLUDING THE MAGNIFICENT ANDROMEDA GALAXY (PICTURED HERE) AND SOME OF THE MOST BEAUTIFUL NEBULAE OF THE YEAR.

TURN THE PAGE, AND DISCOVER THE 15 BEST ASTROPHOTOGRAPHY TARGETS FOR THE FALL SEASON.

FALL #1

DIFFICULTY: ★☆☆

MESSIER 31
THE ANDROMEDA GALAXY

The Andromeda Galaxy

M31 is by far the easiest galaxy to image for beginners.

Because of its diameter of more than six times that of the Moon, we recommend imaging this target with a small telescope. Using a telescope with a long focal length it fine, but the galaxy will not fit completely in the frame so you will be forced to create a mosaic. The left image was taken with a simple 50mm lens attached on a Canon t3i, with exposures of 3 minutes for 4 hours. The one on the right was taken with the same camera and exposure time, but using a telescope of 800mm in focal length.

Do not underestimate the editing process. The core is very bright and the spiral arms are full of details to reveal!

COOL FACTS

- Thought to be a nebula until the 1920's
- More than six times the size of the Moon
- Will collide with Milky Way in 3.75 billion years to create the Milkomeda galaxy

DESIGNATION	M31
TYPE	Galaxy
CONSTELLATION	Andromeda
MAGNITUDE	3.44
SIZE	178' x 63'

- Suggested minimum focal length: 50mm
- Ideal focal length: 400mm

FINDING YOUR TARGET

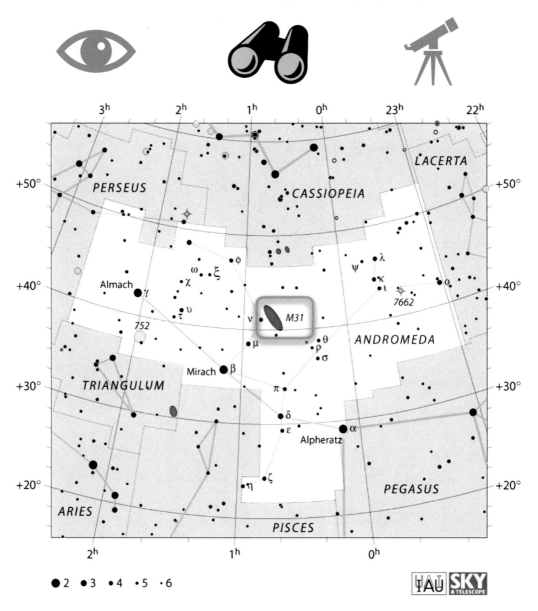

The Andromeda Galaxy is one of the few galaxies visible to the naked eye from Earth in the Northern hemisphere. It is also one of the biggest objects in the night sky.

M31 lies between Pegasus and Cassiopeia. The easiest way to find it is to first locate Pegasus' Great Square. Next, look for the corner closest to Cassiopeia's "W" shape and jump to the first, then hop over to the second star out of that corner. That second star should be the bright Mirach. Finally, do the same double star hop and in a 45 degree angle from that star you will find yourself staring at the brightest galaxy in the Messier catalog! With the naked eye, it will appear as a fuzzy spot next to the star.

Note that viewing this target with binoculars or a telescope will also reveal its satellite galaxy: M110! You may also see M32, but it will most likely be washed out by M31's gases.

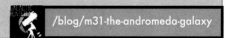
/blog/m31-the-andromeda-galaxy

FALL #2

DIFFICULTY: ★★☆

MESSIER 15
THE GREAT PEGASUS CLUSTER

The Great Pegasus Cluster

This beautiful ball of stars is one of the easiest clusters to photograph.

In order to get the best results, make sure to keep your exposure times low, and to spend about 2 hours on it at least.

For such a busy cluster there is no room for even the slightest star trail, which is why taking short exposure shots of about 30 to 60 seconds will lower the risk dramatically compared to long exposure shots of 3+ minutes.

Lowering the exposure time will also allow you to avoid blowing out the bright center of the cluster.

COOL FACTS

- One of the oldest known globular clusters in our galaxy
- 360,000 times brighter than the sun
- Contains more than 100,000 stars

DESIGNATION	M15
TYPE	Cluster
CONSTELLATION	Pegasus
MAGNITUDE	6.2
SIZE	18'

- Suggested minimum focal length: 300mm
- Ideal focal length: 1000mm

FINDING YOUR TARGET

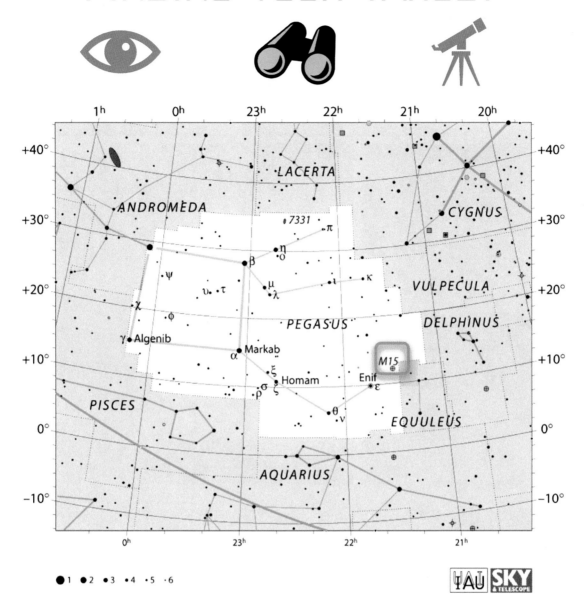

Messier 15 is a very bright, colorful and beautiful globular cluster that is easily seen with any instrument. You can also see M15 with the naked eye from a dark zone, and will appear as a tiny, fuzzy ball of faint light.

To find this cluster, simply locate the Pegasus square again, just like you did when looking for the Andromeda galaxy. This time, start from the opposite corner and follow the four brightest stars that form the neck and head of Pegasus. Your target will be just beyond the last star, Enif, to the right.

Make sure to take a peek through your eyepiece before attaching the camera, to have a good look at this wonderful deep sky object.

FALL #3

DIFFICULTY: ★★☆

IC 1805
THE HEART

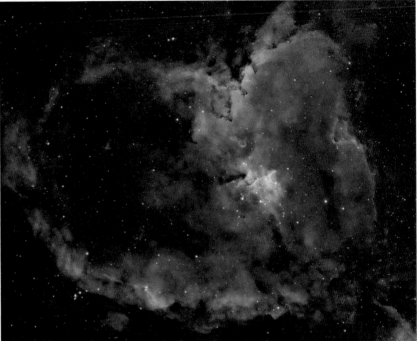

The Heart Nebula

IC 1805 is a huge but faint target with a bright open cluster at its core (Melotte 15).

The best way to image this nebula is to use a telescope with a short focal length (400mm or lower), unless you would prefer to focus on a specific part of the Heart with a bigger telescope.

The most common way to image the Heart is to also include its partner, the Soul nebula (see #4), in the same frame. You can achieve that by either capturing both nebulae individually and composing a mosaic, or by using a DSLR camera on a star tracker. You can also pair a full frame camera with a small telescope to fit the two just right!

COOL FACTS

- About 5 times the size of the Moon
- Most stars are formed in the Heart's center
- Also called the "Running Dog" nebula

DESIGNATION	IC 1805
TYPE	Nebula
CONSTELLATION	Cassiopeia
MAGNITUDE	18.3
SIZE	150' x 150'

- Suggested minimum focal length: 135mm
- Ideal focal length: 650mm

FINDING YOUR TARGET

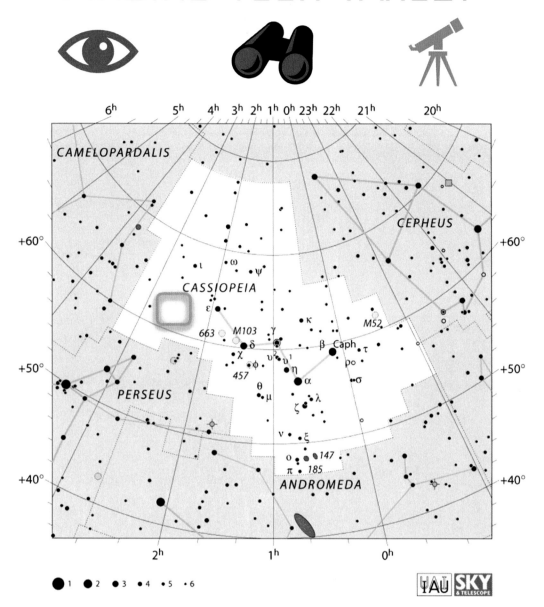

The Heart Nebula, due to the outer parts being washed out by interstellar dust, is not visible to the naked eye or with binoculars. Under the darkest possible skies, binoculars may reveal a tiny, fuzzy cluster at the center of the heart.

IC 1805 is located just to the left of Cassiopeia's "W" shape, and is really not difficult to find due to its size.

If you are unsure of its location, point a DSLR camera with a 50mm lens or so towards Cassiopeia's left side and take a long exposure photo of the general area. The image will reveal some red nebulosity from the Heart nebula and its neighbor, then all there is left to do is to recenter. Re-adjust properly if you plan to do some wide field imaging.

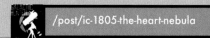
/post/ic-1805-the-heart-nebula

FALL #4

DIFFICULTY: ★★☆

IC 1848
THE SOUL

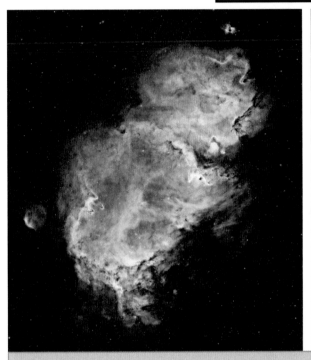

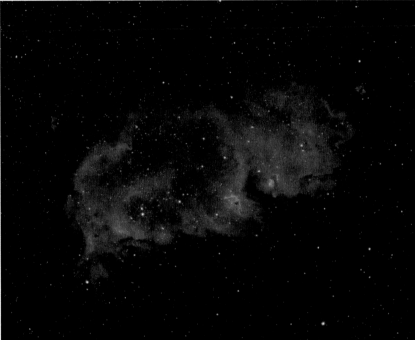

The Soul Nebula

Next to the Heart nebula is the Soul nebula. Because of its size, we recommend imaging this target as you would with the Heart, and use a small telescope or a telephoto camera lens.

A color camera will show an almost completely red nebula, but imaging this same target with narrowband filters will display other colorful gases as well. A comparison can be seen above, with the left image being combined as the Hubble Palette using S, H, and O filters.

Depending on how dark the sky is at your desired location, we would recommend at least 4 hours for either target. You may increase that to 10+ hours if you truly want to capture even the faintest details of this beautiful deep sky object.

COOL FACTS

- Also known as the Embryo nebula
- Two large clouds connected by gas
- Home of the Radio Source W5

DESIGNATION	IC 1848
TYPE	Nebula
CONSTELLATION	Cassiopeia
MAGNITUDE	6.5
SIZE	150' x 75'

- Suggested minimum focal length: 135mm
- Ideal focal length: 650mm

FINDING YOUR TARGET

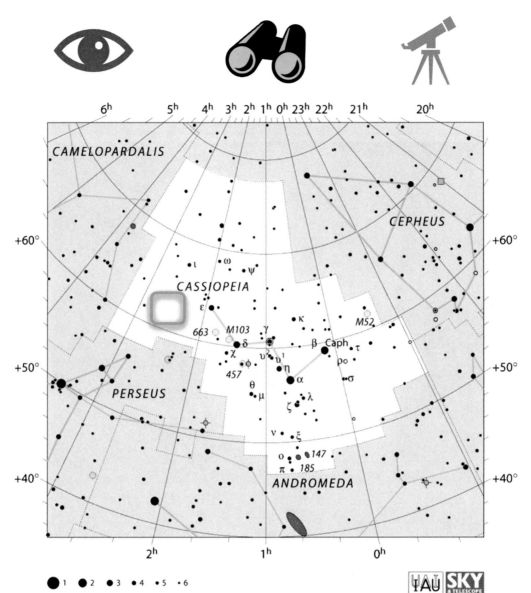

The Soul nebula is located right next to the Heart nebula, so if you find one through a telescope, you most likely will have no problem seeing the other nearby. You can also use the famous Double Cluster in Perseus (see the yellow dot above the "PERSEUS" text) and just go up until spotting the nebulae.

The gases from IC 1848 are noticeably brighter than those of its neighbor, and it has a few barely visible clusters here and there. IC 1848 is one of them, and is located in the body of the Soul. You may also recognize clusters CR 34, CR 632, and CR 634 which are all situated in the head of IC 1848.

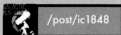
/post/ic1848

FALL #5

DIFFICULTY: ★☆☆

MESSIER 33
THE TRIANGULUM GALAXY

The Triangulum Galaxy

M33 is the second apparent largest, and brightest galaxy to photograph in the night sky.

The Triangulum galaxy is simple to photograph, and good result can be achieved with just 3 hours of exposure time away from light pollution. We recommend between 3 and 6 minutes for each exposure in order to capture faint details, including the huge NGC 604 nebula in one of the spiral arms of M33.

Messier 33 can also be photographed using a simple DSLR camera and lens, as seen on the left image with M31 on the right side. The lens used was 50mm at f/1.8, mounted on a sky tracker for a series of 3 minute exposures. Do you see the Triangulum galaxy at the bottom left?

COOL FACTS

- Large nebula NGC 604 lies in M33
- Will be absorbed by Milkomeda in 2.5 billion years
- One of the most distant naked eye object

DESIGNATION	M33
TYPE	Galaxy
CONSTELLATION	Triangulum
MAGNITUDE	5.72
SIZE	73' x 45'

- Suggested minimum focal length: 50mm
- Ideal focal length: 800mm

FINDING YOUR TARGET

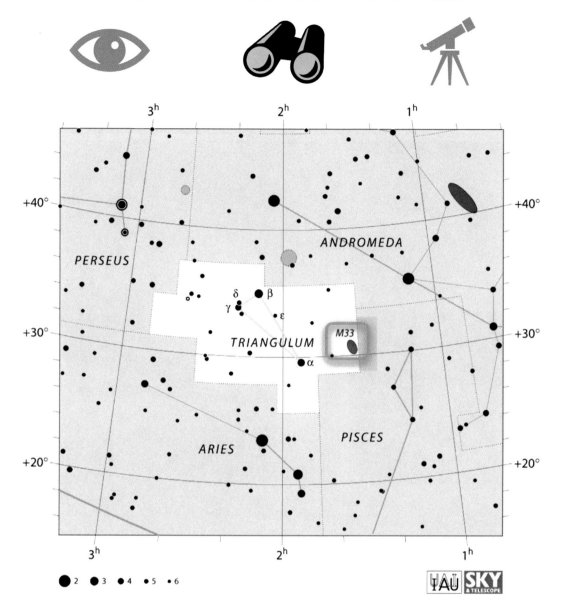

Like the Andromeda galaxy, the Triangulum galaxy can be visible with the naked eye in extremely dark skies far from any light pollution.

The easiest way to spot Messier 33 is to first find M31 (see how in #1 of this Fall section). Once you have M31 in your sight, locate the nearest bright star Mirach. Match the distance between the two, and apply it to the opposite side of this bright star as if its a straight line to spot M33.

Like we said earlier, the Andromeda galaxy is doomed to crash with our own, the Milky Way. M33's fate is no better. The Triangulum galaxy will get stuck in the gravitational pull of the impact and orbit the new Milkomeda until finally crashing into it. In the end, our Milky Way, the Andromeda galaxy, and the Triangulum galaxy... will become one.

/blog/m33-the-triangulum-galaxy

FALL #6

DIFFICULTY: ★☆☆

NGC 7293
THE EYE OF GOD

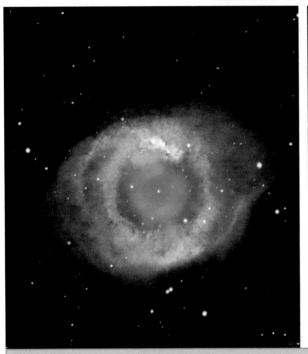
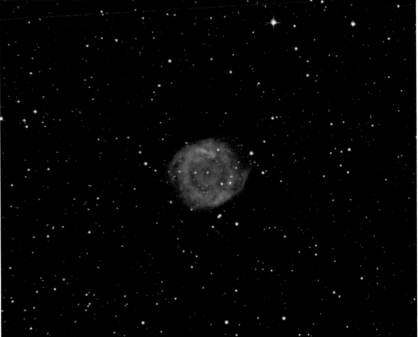

The Helix Nebula

The Helix nebula, commonly known as the Eye of God, or even the Eye of Sauron, looks like a larger and brighter version of M57, the Ring nebula.

This planetary nebula is a very popular target for amateur astrophotographers, not only because it has cool nicknames but it is very easy to capture.

Spending two hours on this target will give you the beautiful, colorful nebula as seen on the image on the right. Adding several more hours will allow you to capture the fainter, outer gases being expelled, as pictured on the left. Notice how the Helix is very similar to the Ring nebula.

COOL FACTS

- First planetary nebula discovered to have cometary knots (about 20,000)
- Expands at a rate of 40km/s
- About half the size of the full moon

DESIGNATION	NGC 7293
TYPE	Nebula
CONSTELLATION	Aquarius
MAGNITUDE	7.6
SIZE	25'

- Suggested minimum focal length: 300mm
- Ideal focal length: 800mm

FINDING YOUR TARGET

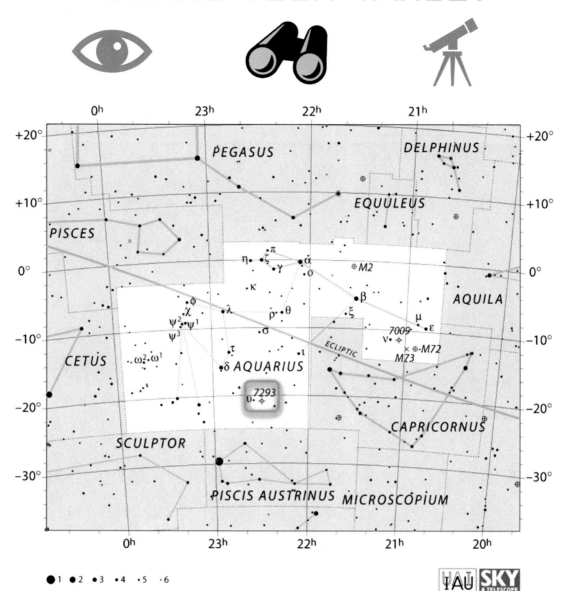

The Helix nebula can be found in the Aquarius constellation, in an area devoid of bright stars.

Even though the Helix is the brightest planetary nebula in the night sky, it is not that easy to observe. Because of its size, the nebula has a low surface brightness, making it difficult to spot through a large telescope. The key is to use binoculars or a low power telescope so that the light from the Helix can be concentrated in one obvious spot in the wide visible sky. NGC 7293 can also be seen with the unaided eye, as long as your vision is excellent and no light pollution is present from your observing site.

Keep in mind that from any place in the United States, the Eye of God does not rise very high in the sky, and disappears after a few weeks. Make sure to capture it before it is gone or you will have no choice but to wait an entire year to meet it again.

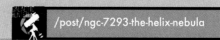
/post/ngc-7293-the-helix-nebula

FALL #7

DIFFICULTY: ★★★

IC 1396
THE ELEPHANT'S TRUNK

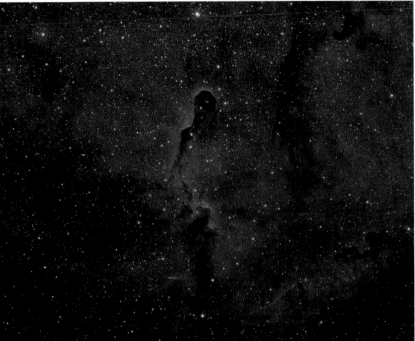

The Elephant's Trunk nebula

IC 1396 is a huge emission nebula. The "Elephant Trunk" is actually a small area located on the Western side of the entire object.

You can see the trunk on both photos above, with a cropped version on the right.

Note that imaging the entire emission nebula with a DSLR is fairly easy, but you will need long hours of exposure time to capture the many details all around and within the object. A color camera will show an almost all red target.

You may also use a very wide telescope in order to capture the dark dust lanes all around.

COOL FACTS

- Trunk is over 20 light-years long
- IC 1396A is the designation for the Elephant's Trunk
- Surrounded by interstellar dust lanes

DESIGNATION	IC 1396
TYPE	Nebula
CONSTELLATION	Cepheus
MAGNITUDE	3.5
SIZE	170' x 140'

- Suggested minimum focal length: 85mm
- Ideal focal length: 400mm

FINDING YOUR TARGET

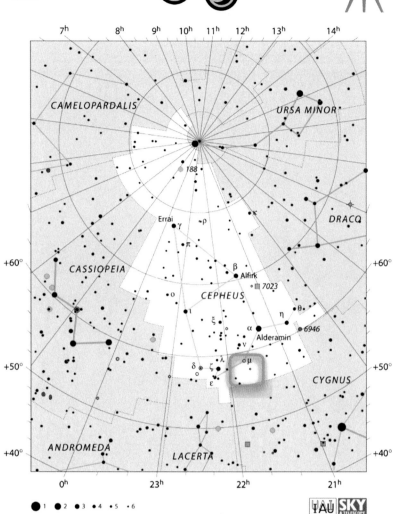

With a magnitude of 3.5, IC 1396 can be seen with the naked eye in very dark skies, and will appear as a gray, very faint blurry patch when looking at it with averted vision.

This huge emission nebula is located in the high constellation of Cepheus, and can easily be found by looking for one of its stars: Mu Cephei, or: "Hershel's Garnet Star". This is easily done, as Mu Cephei is one of the largest and reddest looking star in the Northern Hemisphere.

Using the map above, Mu Cephei, and therefore IC 1396 itself, can be found a little bit Southwest of the bright star Alderamin. Spotting a luminous red target with your naked eye should not be a problem when looking up to Cepheus.

 /post/ic-1396-the-elephant-s-trunk-nebula-from-the-backyard

FALL #8

DIFFICULTY: ★★★

HICKSON COMPACT GROUP 92
STEPHAN'S QUINTET

Stephan's Quintet

Stephan's Quintet is a group of galaxies, tearing each other up while moving rapidly away from us through space.

It was thought that all five galaxies were close together, until the bottom one (NGC 7320) was discovered to be 8 times closer to us than the rest. It was then considered a foreground object rather than part of the group, which is made of NGC 7318A, NGC 7318B, NGC 7319 and NGC 7317.

On the picture to the right, you can see a nearby galaxy, NGC 7331, which is likely to fit your frame if using a telescope under 1000mm of focal length. Properly center the object before shooting! Because of its magnitude and size, Stephan's Quintet is a pretty challenging target.

COOL FACTS

- NGC 7320 is 8 times closer to us than the four other galaxies
- All four galaxies will likely merge with each other

DESIGNATIONS	HCG 92
TYPE	Group of Galaxies
CONSTELLATION	Pegasus
MAGNITUDE	14
SIZE	3'.5

- Suggested minimum focal length: 650mm
- Ideal focal length: 1200mm

FINDING YOUR TARGET

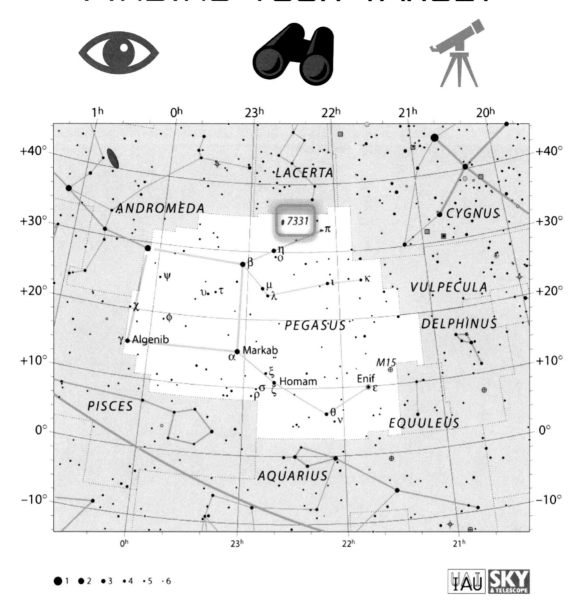

For this target, we are back to the constellation of the flying horse: Pegasus.

We can once again use the main square to guide us to our target. Stephan's Quintet is located near the front legs of the horse and star hop once from the corner linking the front legs towards the next bright star to the East. Then, your target will be a little to the northeast.

Trying to spot Stephan's Quintet with the naked eye or even binoculars would be a waste of time. The group only appears within a 3.5' area in the sky and has a faint 14 magnitude, the only way to see this target is through a telescope. Using an 8" telescope, you might only see a fuzzy elongated patch, which would be the brightest of them: NGC 7318b.

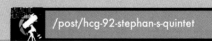

/post/hcg-92-stephan-s-quintet

FALL #9

DIFFICULTY: ★★☆

NGC 7635
THE BUBBLE

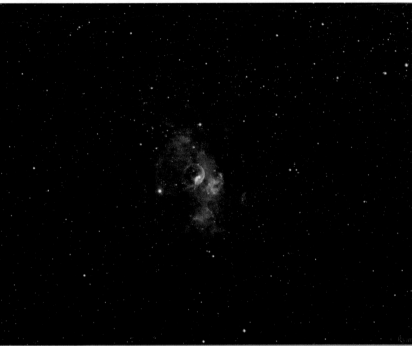

The Bubble Nebula

NGC 7635 is not impressive when looking at it through an eyepiece, but is a magnificent target to photograph.

The bubble itself is very small, but the diffuse gas all around is what makes the beautiful scenery.

Because of how diffuse the gases are, long exposure shots, such as 6 minutes, are recommended. The bubble is not difficult to photograph in three or four hours, but the photons from the faint gases all around will need more time.

You may want to spend a couple nights on this target to obtain a result you really love.

COOL FACTS

- Discovered in 1787
- Has a radius of 3 to 5 light-years
- Created from a huge young central star's stellar wind

DESIGNATION	NGC 7635
TYPE	Nebula
CONSTELLATION	Cassiopeia
MAGNITUDE	10.0
SIZE	16' x 9'

- Suggested minimum focal length: 135mm
- Ideal focal length: 650mm

FINDING YOUR TARGET

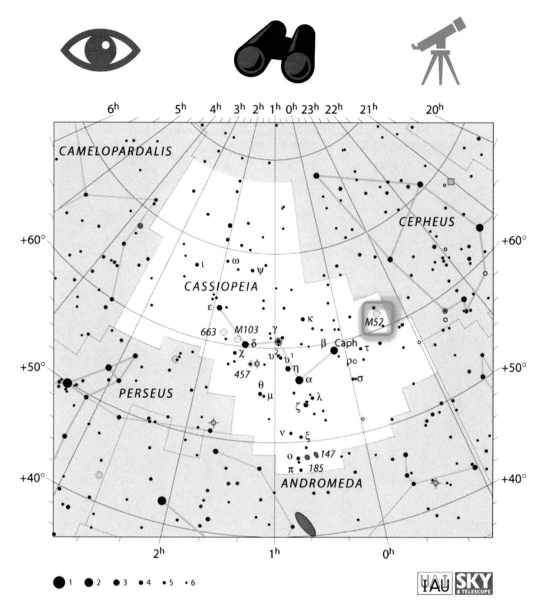

The Bubble nebula is located only 0.5 degrees southwest of Messier 52, an open cluster also in the constellation of Cassiopeia.

The easiest way to find this target is to simply follow the right line of the "W" shape to the northeast.

Again, this is not a visible object with the naked eye or binoculars, and using averted vision is recommended even when looking through a telescope eyepiece because of how diffuse the nebulosity is. Filters may also help, but this is far from being a visually stunning target when looking at it directly.

If you are using a small telescope to capture the Bubble nebula, you will also be able to include the open cluster Messier 52 in your frame, as well as the colorful Lobster Claw nebula (Sh2-157).

/post/ngc-7635

FALL #10

DIFFICULTY: ★★★

NGC 7023
THE IRIS

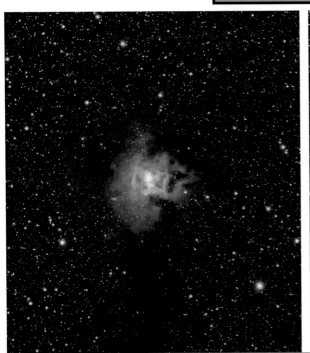

The Iris Nebula

If there was a place to call heaven in space, this would be it. This magnificent, blue reflection nebula has a bright young star in its center, and gradually darkens as it is surrounded by clouds of interstellar dust.

The reason why this target has a 3-star difficulty is because of the processing of the images. Interstellar dust is very difficult to work with and requires special editing techniques. The fact that the central star is so bright and that the nebula itself has an enormous amount of details does not help either. If you feel like you have the processing skills needed for this wonderful nebula, go ahead and capture it, but be ready to spend several hours processing it.

COOL FACTS

- NGC 7023's blue petals are six light-years across
- Surrounded by huge clouds of interstellar dust

DESIGNATION	NGC 7023
TYPE	Nebula
CONSTELLATION	Cepehus
MAGNITUDE	6.8
SIZE	18' x 18'

- Suggested minimum focal length: 135mm
- Ideal focal length: 650mm

FINDING YOUR TARGET

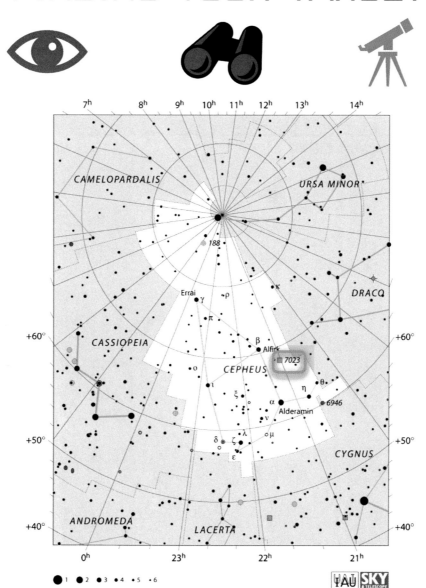

You can find the beautiful Iris nebula in the Cepheus constellation.

If you are having trouble finding Cepheus, first locate the North star. The constellation we need is the one just to the south of Polaris. Once again, look for the bright star Alderamin, and simply go north until reaching another bright star (Alfirk). NGC 7023 will be a little bit to the southeast.

Just like the bubble nebula, viewing the Iris with binoculars is a nearly impossible task. We recommend traveling to a dark site (Bortle 1 to Bortle 3) to capture or observe this object.

/post/ngc-7023-the-iris-nebula

FALL #11

DIFFICULTY: ★★★

IC 5146
THE COCOON

The Cocoon Nebula

IC 5146's shape is similar to the Iris nebula, but this target would be hell instead of heaven.

Like the Iris nebula, the Cocoon nebula is surrounded by interstellar dust lanes and looks like it has a long tail on one side!

Do not center this target in your camera unless you are using a large instrument, because capturing its dark tail makes the overall image more impressive than a close up.

Just as the previous target, the Cocoon nebula earns its 3-star difficulty because of the processing skills required to show the dust lanes.

COOL FACTS

- Central star lights up the nebula
- Has interstellar dust lanes forming a tail, called Barnard 168
- IC 5146 is a compact star forming region

DESIGNATION	IC 5146
TYPE	Nebula
CONSTELLATION	Cygnus
MAGNITUDE	7.2
SIZE	12'

- Suggested minimum focal length: 135mm
- Ideal focal length: 400mm

FINDING YOUR TARGET

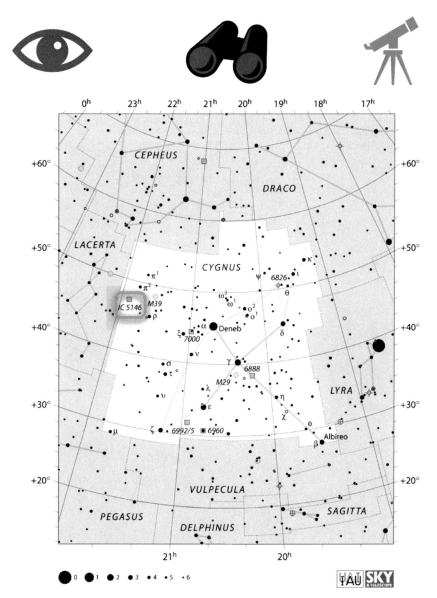

The Cocoon nebula is visually interesting if observed from a very dark zone, mostly because instruments need really good skies to reveal the dark lanes next to the nebula. If you do not have a large instrument or would prefer to use a small telescope, you will still be able to contemplate the cocoon as a glowing, diffuse circular object.

Locating the Cocoon is not an easy task, it lies in the high flying constellation of Cygnus, but is closer to Lacerta's stars. It is near M39, a not-so impressive cluster which is also not easy to find. The best way to land on IC 5146 manually is to draw an imaginary line between the very bright star Deneb and 4 Lac, the star in Lacerta closest to Cygnus. Your target will be about 3/4 of the way from Deneb.

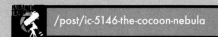

/post/ic-5146-the-cocoon-nebula

FALL #12

MESSIER 74
THE PHANTOM

DIFFICULTY: ★★☆

The Phantom Galaxy

This spiral galaxy, lying 30 million light-years away from us, has a relatively similar size to our Milky Way.

The Phantom galaxy is difficult to observe visually, but easy to photograph and process. It is very similar to M33, the Triangulum galaxy, with a lot of bright stars in the spiral arms.

In photographs, the core will appear white or yellow, while the arms will have a blue or purple tint depending on how it is processed. You will also see some red in the spiral arms if your camera allows it.

COOL FACTS

- Discovered in 1780
- Was initially cataloged as a cluster by John Herschel
- Part of the M74 Group of 5 to 7 galaxies

DESIGNATION	M74
TYPE	Galaxy
CONSTELLATION	Pisces
MAGNITUDE	10.0
SIZE	10'.5 x 9'.5

- Suggested minimum focal length: 350mm
- Ideal focal length: 1000mm

FINDING YOUR TARGET

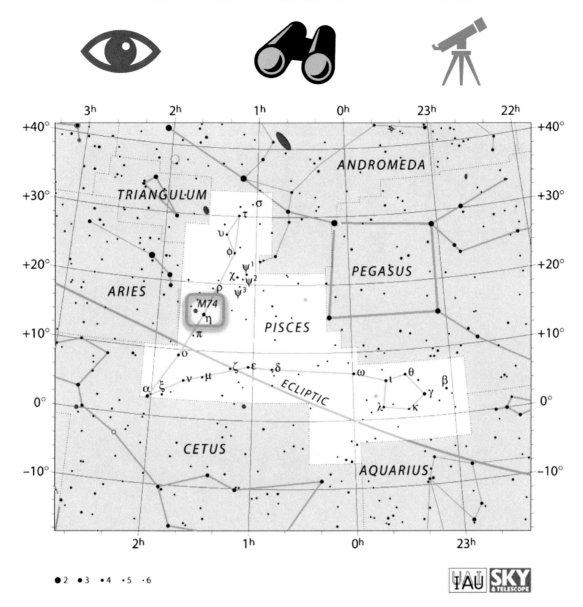

Messier 74, due to its very low surface brightness, is one of the most difficult Messier objects to observe using large binoculars. It will look like a small blurry patch even in an extremely dark site without any light pollution.

Observing this object in a telescope would be the best option, as you will be able to resolve its bright core and some faint surrounding gas without too much difficulty.

The Phantom galaxy lies in the Pisces constellation, right next to the constellation of the flying horse Pegasus. To find it, first locate the brightest star in Pisces, Kullat Nunu. Your target will be 1.5 degrees northeast from that star.

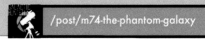
/post/m74-the-phantom-galaxy

FALL
#13

DIFFICULTY: ★★☆

MESSIER 92
CLUSTER IN HERCULES

Messier 92

M92 is one of the oldest clusters known, at 14.2 billion years old.

This globular cluster has a similar size and magnitude as M15, and is a great target for beginner astrophotographers.

Make sure to follow our advice for clusters on exposure times and perfect guiding, so your final image will look crisp and bright.

You will not need to spend a long time on M92, so we suggest dedicating two hours on it after imaging another target!

COOL FACTS

- As old as the Universe
- Discovered in 1777
- Approaching us at a speed of 112km/s

DESIGNATION	M92
TYPE	Cluster
CONSTELLATION	Hercules
MAGNITUDE	6.3
SIZE	14'

- Suggested minimum focal length: 250mm
- Ideal focal length: 1000mm

FINDING YOUR TARGET

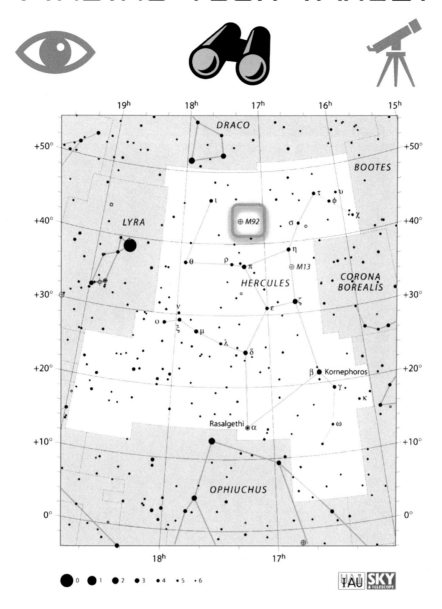

Located in the constellation of Hercules, the globular cluster can be found northwest of M13, preceding the left foot of Hercules.

Barely visible to the naked eye, M92 can be best seen with small binoculars, but will look like a fuzzy star. A small telescope will reveal the cluster with several stars around a bright core. The use of a bigger instrument will allow you to resolve much more stars that would be invisible at lower aperture.

For comparison, M92 has a magnitude of 6.3 and a size of 14', while the more-popular cluster Messier 13 has a magnitude of 6.2 and a size of 18'. Take a look at each and see which one is your favorite!

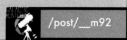

FALL #14

DIFFICULTY: ★★★

IC 405
THE FLAMING STAR

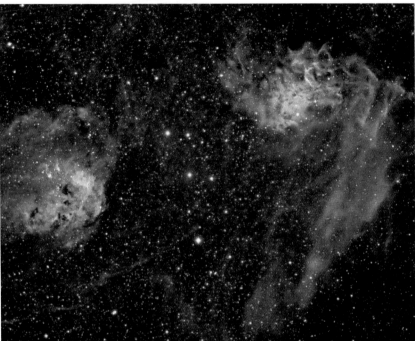

The Flaming Star Nebula

IC 405 is a beautiful emission/reflection nebula, composed of gases surrounding the very hot blue star AE Aurigae.

This target is great for astrophotographers who are used to easy Messier objects and would like to test their skills on a more complex nebula.

The Flaming Star nebula gets its 3-star difficulty rating due to its faint details and red color all over. This can be challenging to process if imaging from a light polluted zone or without enough exposure time with a DSLR camera. On the left image, you can see a close up shot of the Flaming Star Nebula. The right image shows both IC 405 and a nearby nebula IC 410, also called the Tadpoles nebula.

COOL FACTS

- Spans about 5 light-years
- Proper motion of the central star can be traced back to Orion's Belt
- Will fade away in 20,000 years

DESIGNATION	IC 405
TYPE	Nebula
CONSTELLATION	Auriga
MAGNITUDE	6.0
SIZE	37' x19'

- Suggested minimum focal length: 135mm
- Ideal focal length: 650mm

FINDING YOUR TARGET

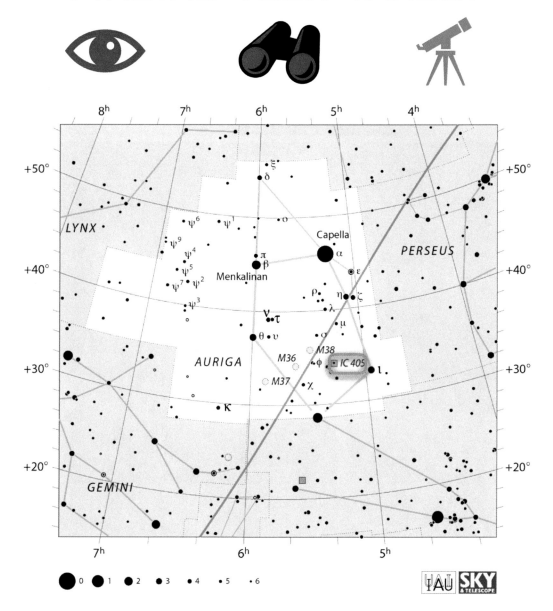

The Flaming Star nebula can be found in the Auriga constellation. It lies near two open clusters, Messier 38 and Messier 36.

IC 405 can not be seen with the naked eye, and only its bright star is visible through binoculars. The nebulosity itself is so faint that it is challenging to see even through a telescope.

The Flaming Star nebula looks like it has blurry flames coming out of the star when seen through a good telescope. As a challenge, try to spot the nearby nebula IC 410, but know that it is almost impossible to see since it is so faint.

 /post/the-flaming-star-nebula

FALL #15

DIFFICULTY: ★☆☆

NGC 281
PACMAN

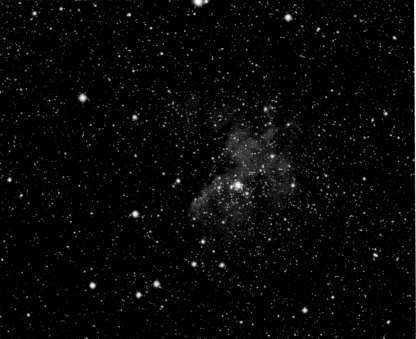

The Pacman Nebula

NGC 281 is an interesting emission nebula floating in the Milky Way's Perseus Spiral Arm.

The Pacman nebula got its name because it looks like the famous titular video game character when seen in wide field. The nebula has a young star cluster, IC 1590, within its star forming region, which lights up the gases around it.

With a magnitude of +7.0, this target is great to photograph and easy to process. Note that like most other deep sky objects, narrowband filters will show much more detail within the gases. It is still a great target for basic DSLR photography though, as shown in the example images above.

COOL FACTS

- Discovered in 1883
- Resembles a video game character
- Star cluster IC 1590 within the gases

DESIGNATION	NGC 281
TYPE	Nebula
CONSTELLATION	Cassiopeia
MAGNITUDE	7.0
SIZE	20' x 30'

- Suggested minimum focal length: 135mm
- Ideal focal length: 800mm

FINDING YOUR TARGET

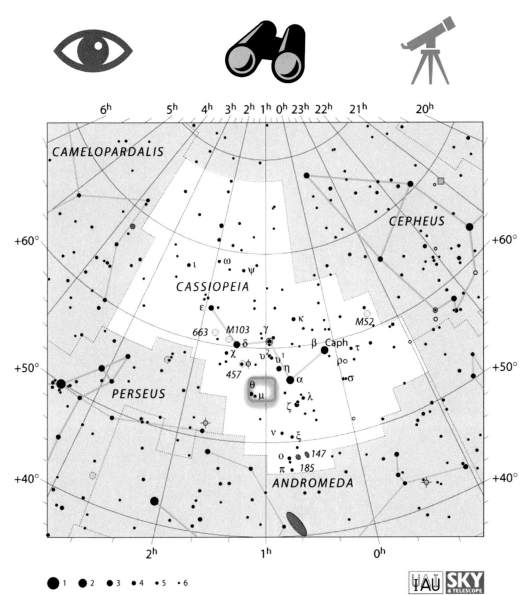

Despite its brightness, the Pacman nebula is not visible with the naked eye, but can be seen through binoculars in extremely dark skies. It is easily seen through a telescope, and will look like a gray cloud with a tiny star in the middle.

NGC 281 can be found in Cassiopeia, not far from the huge Andromeda galaxy. The target is located just South of the star Shedir (bottom right star of the "W" shape).

Other popular deep sky objects lie near the Pacman Nebula, like the Heart and Soul Nebulae, the open clusters M52 and M103, the Bubble Nebula, the Lobster Claw Nebula, the Double Cluster in Perseus, and more.

/post/ngc-281-the-pacman-nebula

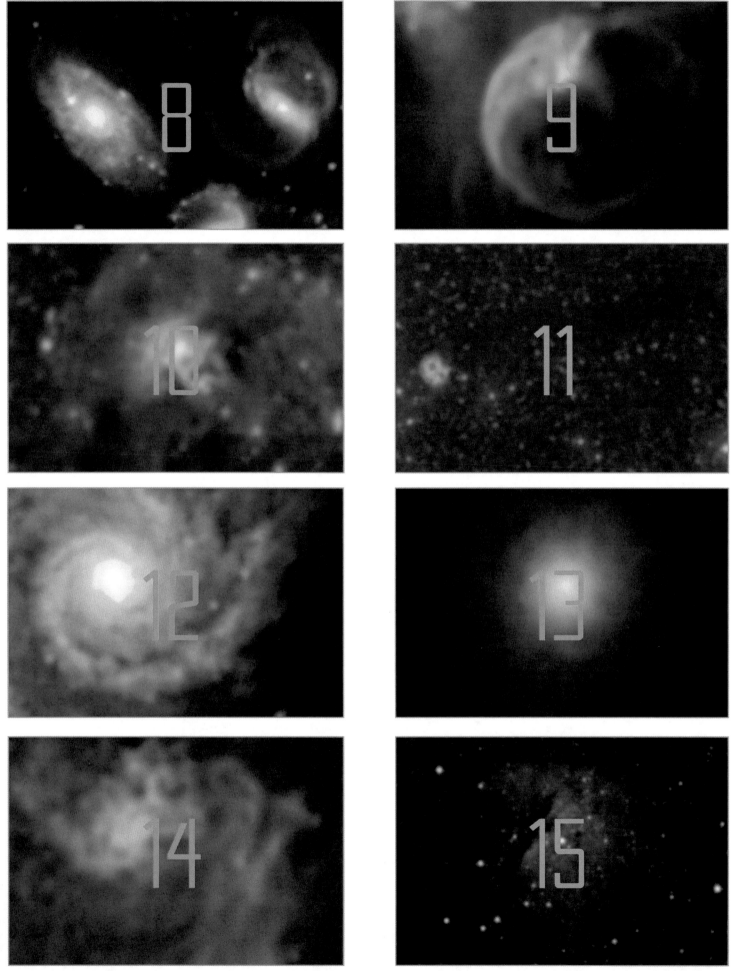

THE 15 BEST TARGETS of... WIN

TER

Winter is here! And it does not disappoint. Two of the easiest and most impressive targets for amateur astrophotographers can be captured during this season: the famous Orion Nebula, and the majestic Pleiades cluster.

You will also be able to photograph the Christmas Tree Nebula, right on time for the holidays.

WINTER #1

DIFFICULTY: ★★☆

MESSIER 1
THE CRAB

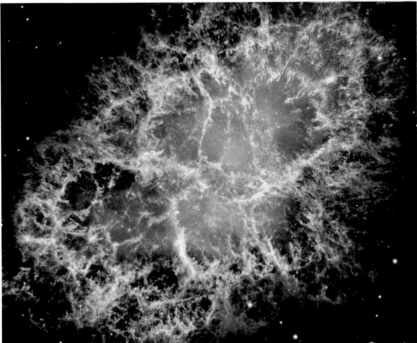

The Crab Nebula

The first Messier object, the Crab nebula, is one of the most famous supernova nebulae in the night sky.

The Crab was born in 1054 when a star died, and the event was so bright that it was visible during the day for 23 days, and also to the naked eye at night for almost two years!

Although rather small, this is a great target for astrophotography, because it is colorful and full of details. It is also pretty iconic, as it is after this discovery, mistaken for a comet, that Charles Messier decided to start building his catalog of "non-comet deep sky objects".

The photo on the left is only 1 hour and a half of total exposure using a stock DSLR camera.

COOL FACTS

- Supernova reported in China in 1054 AD
- Only supernova remnant in the Messier catalog
- Expands at a rate of 1,500 km/s

DESIGNATION	M1
TYPE	Nebula
CONSTELLATION	Taurus
MAGNITUDE	8.4
SIZE	420" x 290"

- Suggested minimum focal length: 650mm
- Ideal focal length: 1200mm

FINDING YOUR TARGET

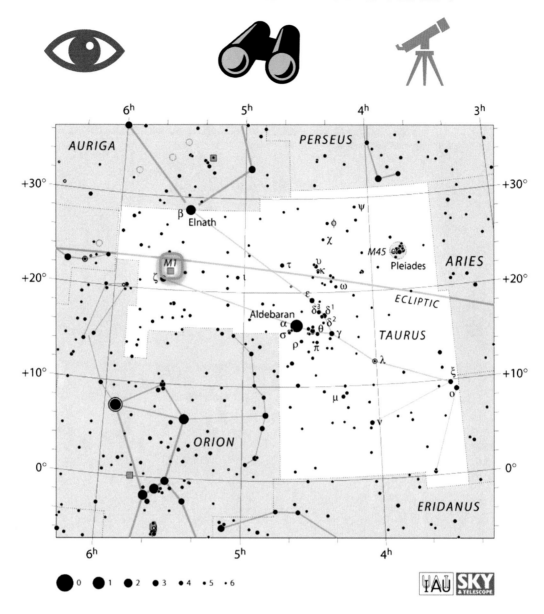

M1 is located in the constellation of Taurus, the bull. It is pretty easy to find because it is close to a bright star in the constellation, Zeta Tauri.

The Crab nebula is not visible with the naked eye, at least not anymore. It is very difficult to spot through binoculars, although possible if you manage to spot the difference between the nebula and the nearby stars. Messier 1 is easily visible with any telescope, but a 16" or bigger instrument will reveal details within the gases.

If you have the chance to use a large telescope, you might be able to spot the Crab Pulsar, a magnitude 16 pulsar that spins more than 30 times per second. One of the most impressive astrophotography time-lapses was done on the Crab nebula. Be sure to look it up online to see how the gasses in the object interact over the years.

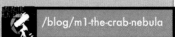
/blog/m1-the-crab-nebula

WINTER #2

DIFFICULTY: ★★☆

MESSIER 35 CLUSTER IN GEMINI

Messier 35

M35 is a large, 110 million year old cluster located 2,800 light-years away.

Another cluster, NGC 2158, is next to Messier 35, and the both of them together make for a beautiful photo full of stars.

The only tricky part about photographing this target with its cluster neighbor is to properly center the camera so you can capture them both without cutting either one off.

Make sure your tracking and guiding are on point so that the stars within the clusters don't drift over each other.

COOL FACTS

- Discovered around 1750
- The only Messier Object in Gemini
- About the same size as the full moon

DESIGNATION	M35
TYPE	Cluster
CONSTELLATION	Gemini
MAGNITUDE	5.3
SIZE	28'

- Suggested minimum focal length: 200mm
- Ideal focal length: 800mm

FINDING YOUR TARGET

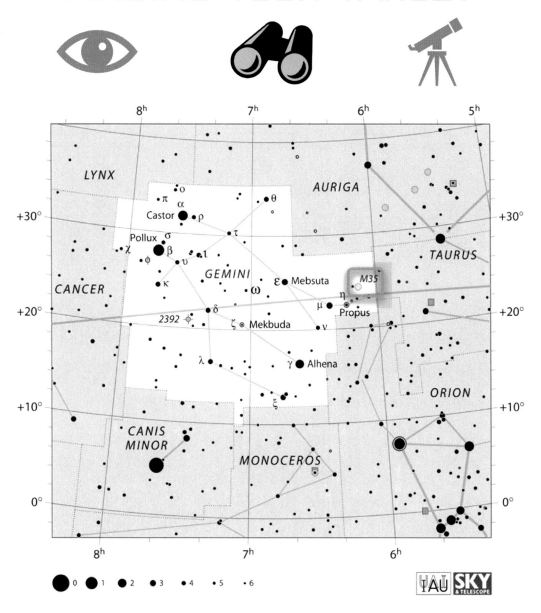

Messier 35 can be found in the Gemini constellation. The cluster can easily be found by locating the nearest bright star: Capella, from the Taurus constellation. Starting from Alhena, a bright star in the constellation of Gemini, move in a straight line to Capella and the line will lead you directly to M35. You should land on the cluster a little more than halfway through.

The cluster can be seen with the naked eye under the best possible conditions, and can easily observed through binoculars as well as any size telescope.

If your goal includes seeing the neighboring cluster NGC 2158, a larger telescope will be needed. That second cluster can be found just 15 arc minutes to the southwest of M35.

/blog/messier35

WINTER #3

DIFFICULTY: ★☆☆

NGC 1499
CALIFORNIA

The California Nebula

This long cloud of hydrogen gas is NGC 1499, or, as its shape implies, the California nebula!

This emission nebula gets its glow from Xi Persei, which you can see shine bright on both pictures above.

The California nebula, despite being a famous target for astrophotographers, gets a 3-star difficulty rating because it is really hard to spot with a telescope. Getting the right angle to make the entire state fit in the frame is a blind challenge. Make sure to spend at least three hours on this target. You might also want to wait until you purchase a filter before attempting to capture it, or modify your DSLR camera. An easier option would be to capture it wide-field, with M45 in the same frame.

COOL FACTS

- Shaped like the U.S. state of California
- Located in the Orion arm of our galaxy
- Gets its glow from the extremely hot O-class star Xi Persi

DESIGNATION	NGC 1499
TYPE	Nebula
CONSTELLATION	Perseus
MAGNITUDE	6.0
SIZE	2.5°

- Suggested minimum focal length: 135mm
- Ideal focal length: 650mm

FINDING YOUR TARGET

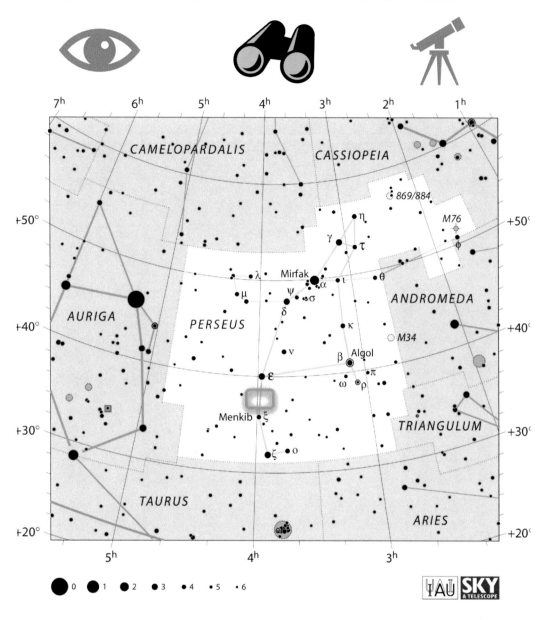

The California nebula can be found in the Perseus constellation, just north of the star Menkib.

Because it is filled with Hydrogen gas, the California nebula is extremely difficult to see visually. The reason why all three visibility icons are green and not red, is because the nebula can be seen easily by using a filter, such as: OIII, UHC, or even better, a H-Beta filter.

Holding any of those filters to your eye and looking up will allow you to see this target without any instrument, as long as you are far from light pollution and have perfect eyesight. Know that NGC 1499 looks like the California state in photos, but do not expect to see the iconic shape while viewing it. This is one of the hardest popular targets to see and looks like a patch of sky that is slightly lighter.

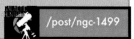
/post/ngc-1499

WINTER #4

DIFFICULTY: ★☆☆

MESSIER 45
THE PLEIADES

The Pleiades

Messier 45, the beautiful Seven Sisters of the night sky, just about the easiest star cluster to photograph!

M45 is huge, extremely bright, and looks amazing in close up shots using a telescope (right) and wide field photography (left, seen with the California nebula). The stars in the Pleiades are blue and glow against the faint nebulosity that they are passing through.

Taking a photo with an exposure of 3 minutes is enough to see the reflection nebulae around the stars. The photo on the right was the result of only 2 hours of total exposure with an 8" reflector telescope and an unmodified DSLR camera from a Bortle 4 site!

COOL FACTS

- First observed by Galileo Galilei
- Moving towards the Orion constellation
- Names of the seven brightest stars making the Seven Sisters are Sterope, Electra, Merope, Maia, Celaeno, Taygeta & Alcyone

DESIGNATION	M45
TYPE	Cluster
CONSTELLATION	Taurus
MAGNITUDE	1.6
SIZE	110'

- Suggested minimum focal length: 50mm
- Ideal focal length: 650mm

FINDING YOUR TARGET

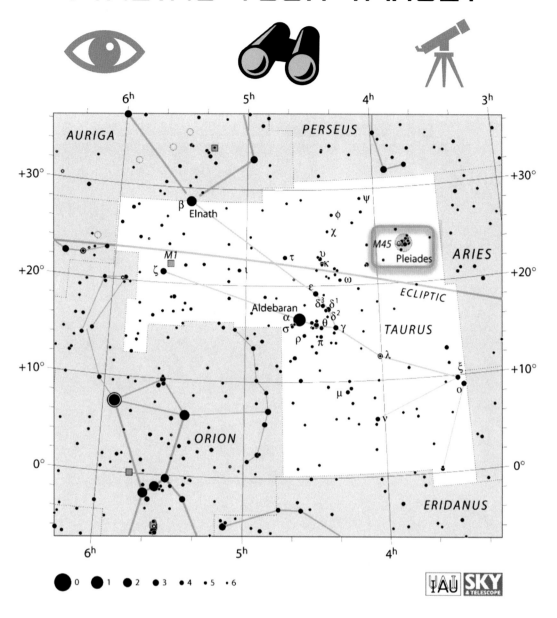

The Pleiades are by far the easiest Deep Sky Object to spot in the night sky. They can be seen with the unaided eye, binoculars, and of course telescopes. Light pollution does not affect the visibility of the main stars of the Seven Sisters, as they can be spotted from a dark site and in the city. With the naked eye, the cluster looks like several bright blue stars linked to one another by small lines. Through binoculars, it looks like a set of big, luminous stars forming the shape of an uneven cross. The nebulosity around the stars may be seen with a large aperture telescope, although it is very faint and difficult to make out.

M45 is located in the Taurus constellation, and is about four times the apparent size of the full moon. The cluster is slowly dispersing and drifting towards Orion, and will officially be part of the hunter constellation in about 250 million years.

 /blog/m45-the-pleiades-star-cluster

WINTER #5

DIFFICULTY: ★★☆

MESSIER 44
THE BEEHIVE

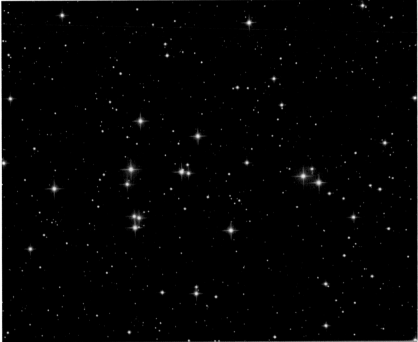

The Beehive Cluster

The Beehive cluster may lack exciting features but it is still a bright and large star cluster.

M44 is the third brightest object in the Messier catalog. In photographs, it appears mostly white and blue, but the cluster also contains a few red and orange stars.

It is possible to make M44 look quite beautiful in photographs depending on the types of star spikes your telescope's spider veins will produce.

As usual, make sure to check your tracking and guiding. Avoid doing very long exposures for each shot so the stars do not get blown up.

COOL FACTS

- Contains about 1,000 stars
- One of the nearest star clusters to Earth
- Also called Praesepe, or The Manger

DESIGNATION	M44
TYPE	Cluster
CONSTELLATION	Cancer
MAGNITUDE	3.7
SIZE	95'

- Suggested minimum focal length: 135mm
- Ideal focal length: 500mm

FINDING YOUR TARGET

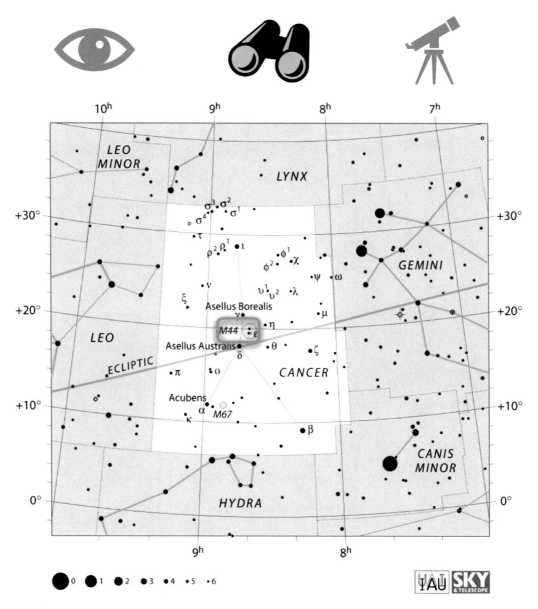

M44 can be found in the faint constellation of the Cancer. Due to its brightness, the Beehive Cluster is visible to the naked eye but is best seen through binoculars or small telescopes. The only worthwhile thing about pointing a large telescope at this target is to observe the hundreds of faint stars that would otherwise be invisible with smaller instruments.

The Beehive is best viewed near the end of Winter through the beginning of Spring, as Cancer rises higher when Winter is ending.

The best way to find the cluster is to ignore the faint stars from Cancer, and instead look for the two brightest stars in Gemini: Pollux and Castor. Draw an imaginary line starting from Castor that goes straight through Pollux and continues. Follow the imaginary line until you reach the Beehive cluster. If unable to do so, M44 also lies close to the Cancer star Asellus Borealis.

/blog/m44-the-beehive-cluster

WINTER #6

DIFFICULTY: ★★☆

MESSIER 95
BARRED SPIRAL IN LEO

Messier 95

Messier 95 is a barred spiral galaxy that is part of the M96 galaxy group also known as the Leo I Group.

Because of its small size and great amount of detail, M95 is best photographed with a large telescope. If using a telescope with a focal length of 800mm or smaller, you can center the view right in between M96 and M95. By doing so, you will be able to capture both of those galaxies and others from the M96 Group, including Messier 105! (More details on #7)

The photo on the right is from NASA, while the one on the left is a cropped image that was taken with a telescope and DSLR camera. We recommend spending at least 3 hours on this target.

COOL FACTS

- Receding from the Milky Way at 778 km/s
- One of the faintest objects in the Messier Catalog
- Contains about 40 billion stars

DESIGNATION	M95
TYPE	Galaxy
CONSTELLATION	Leo
MAGNITUDE	11.4
SIZE	3'.1 x 2'.9

- Suggested minimum focal length: 600mm
- Ideal focal length: 800mm

FINDING YOUR TARGET

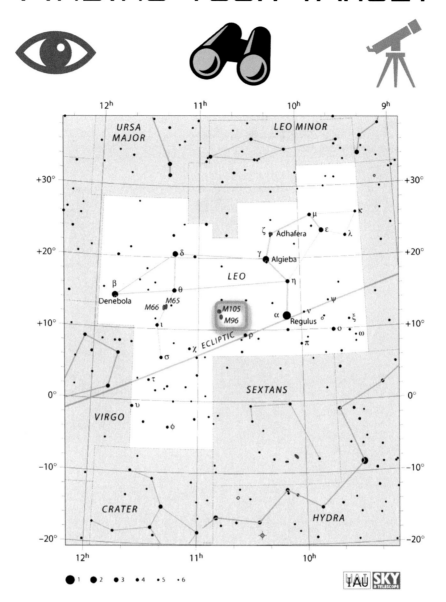

Messier 95 has a magnitude of 11.4, and is impossible to see with the naked eye. It is also extremely difficult to spot with binoculars and you would need very large ones.

M95 is one of the faintest Messier objects in the entire catalog, so the best way to look at this target is through a telescope. Most telescopes will only show the core with faint gas around it, but large instruments under perfectly dark skies will reveal more detail in the spiral arms.

M95 is located in the constellation of Leo, and is the neighbor of several other galaxies, as it is part of the M96 Group. To find them, start from Regulus, the brightest star of the constellation, then make your way in a straight line to Denebola, which is also in Leo. You should cross over the M96 group about one third of the way there.

/blog/m96-group-8-galaxies-in-the-constellation-of-leo

WINTER #7

DIFFICULTY: ★★☆

MESSIER 96
INTERMEDIATE SPIRAL IN LEO

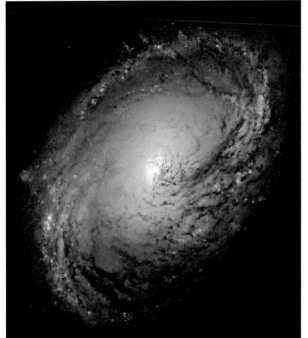

Messier 96

M96 is considered an intermediate spiral galaxy, and just like M95, it is part of the Leo I Group of galaxies. The group is made of more than 12 galaxies, including three Messier objects.

Astronomers found that there was a greater concentration of dust in the right side of the galaxy, and deducted that the right side of M96 was closer to us than the left.

Just like for M95, we recommend photographing this galaxy either by itself with a large instrument, or using a medium size telescope and include as many neighbors as possible (7 galaxies total can be seen on the right photo, M95 on the top right, M96 on the bottom right, and M105 on the left, looking like a bright blob of light next to two other NGC galaxies).

COOL FACTS

- Receding from the Milky Way at 897 km/s
- Brightest and largest member of the M96 Group
- Contains about 100 billion stars

DESIGNATION	M96
TYPE	Galaxy
CONSTELLATION	Leo
MAGNITUDE	10.1
SIZE	7'.6 x 5'.2

- Suggested minimum focal length: 600mm
- Ideal focal length: 800mm

FINDING YOUR TARGET

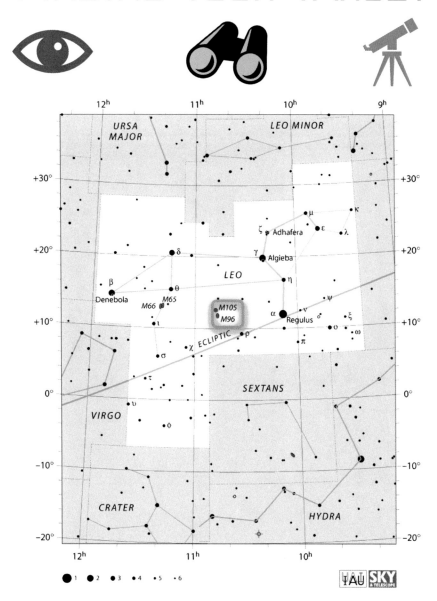

M96 is located right next to M95, in the ribs of the Lion constellation. Simply repeat the steps on how to find the 95th Messier object and you're there! If using a large telescope, you may not be able to see both in the same field of view. Note that M96 is just 40 arc minutes east of M95.

M96's visibility is exactly like M95's. It requires extremely dark skies, far from light pollution, and is only visible through large binoculars or telescopes.

You can compare your framing with another nearby group: The Leo Triplet! The triplet is also in the Leo constellation located about halfway from the M96 Group and Denebola.

To ensure you get as many objects in your image as possible, rotate your camera and keep taking test shots until you can fit the other nearby galaxies in there!

 /blog/m96-group-8-galaxies-in-the-constellation-of-leo

WINTER
#8

DIFFICULTY: ★★☆

MESSIER 37
OPEN CLUSTER IN AURIGA

Open cluster in Auriga

There are three open clusters in the constellation of Auriga, M37 being one of them. It is the largest, brightest, and most populated of the three.

Photographing this object is simple and obtaining great results does not require many hours of exposure. The photo on the right was captured in 45 minutes at ISO 800. The one on the left totals 13 hours. It is still a beautiful cluster of stars to capture, and you might notice a red patch on the left image. This is the oldest planetary nebula every discovered, and can be captured with a HA filter!

As usual with clusters, we decided to give it a two-star rating of the difficulty because the guiding needs to be perfect.

COOL FACTS

▸ Discovered before 1654
▸ Contains about 500 stars, at least 12 of them evolved from red giants
▸ Very close to M36 and M38

DESIGNATION	M37
TYPE	Cluster
CONSTELLATION	Auriga
MAGNITUDE	6.2
SIZE	24'

▸ Suggested minimum focal length: 400mm
▸ Ideal focal length: 800mm

FINDING YOUR TARGET

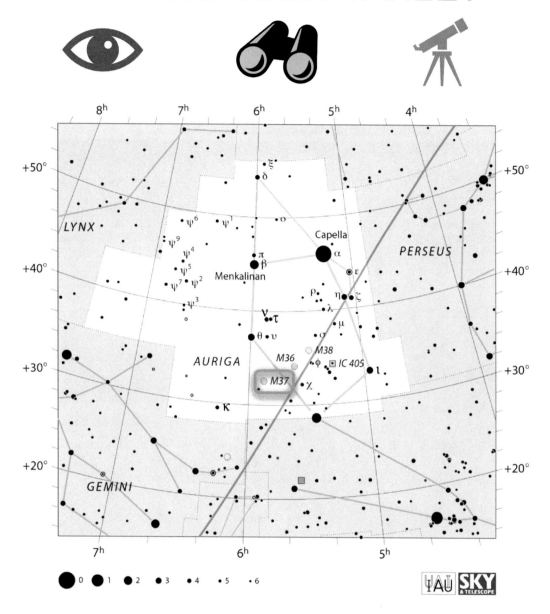

M37 lies in the constellation of Auriga, in between Gemini and Perseus.

To find it, first locate the sixth brightest star in the night sky: Capella. The cluster is on the opposite side of that star towards Gemini, just to the exterior of Auriga's pentagon asterism. Unlike M37, both M36 and M38 are inside the pentagon shape.

Messier 37 can not be resolved with the naked eye, but its brightest stars can easily be seen through binoculars. A small telescope will be able to show you about 20 bright stars, while bigger instruments will reveal hundreds of stars, including the fainter members.

The planetary nebula seen within M37 was discovered only recently. We spent 10 hours with our HA filter to reveal it, and 3 hours of R/G/B. Its name is "IPHASX J055226.2+323724".

/blog/m37-open-cluster-in-auriga

WINTER #9

DIFFICULTY: ★☆☆

MESSIER 42
ORION

The Orion Nebula

Messier 42 is the most popular nebula to photograph! And it is easy to capture!

M42 looks amazing in photographs through both telescopes and DSLR lenses. The photo on the left was taken with a 85mm lens attached to a DSLR camera. You can see a lot of gas surrounding the nebula, and even some of the reds coming from IC 434, the next target on this list! The photo on the right was taken through an 8" telescope with a total exposure time of only 1 hour! M43 (The little ball shaped nebula on the left of the heart shape of M42) is visible, as well as the Running Man nebula (top left). When capturing it, be sure to also take a few short exposure shots for the bright core as well as the Trapezium cluster!

COOL FACTS

- Discovered in 1610
- The nearest stellar nursery to Earth
- The Trapezium is an open cluster that powers the gases all around

DESIGNATION	M42
TYPE	Nebula
CONSTELLATION	Orion
MAGNITUDE	4.0
SIZE	65' x 60'

- Suggested minimum focal length: 50mm
- Ideal focal length: 600mm

FINDING YOUR TARGET

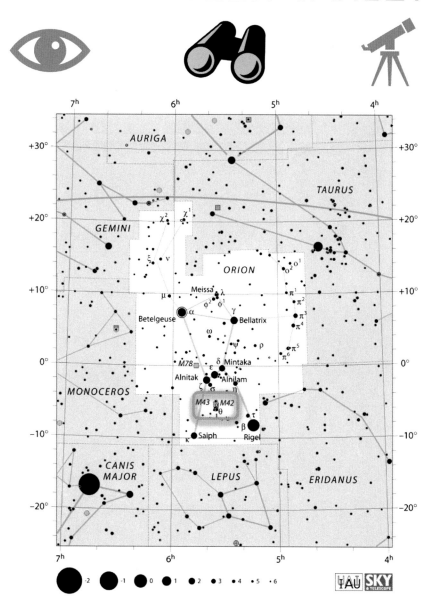

M42 is the easiest nebula to find. It is located in Orion's sword and is obvious to spot as it looks like a star. It lies between other deep sky objects, such as the Horsehead and Flame nebulae, M78, and even the Witch Head nebula.

Messier 42 is one of the brightest nebulae in the sky, and is easily visible with the unaided eye even though it looks just like a regular star. The nebula is an amazing sight through binoculars, as you can make out its shape as well as its bright core. Any size telescope will reveal more of the gases with different shades of gray. You will also be able to spot the four stars that form the famous Trapezium cluster in the core of the nebula.

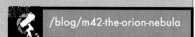

/blog/m42-the-orion-nebula

WINTER
#10

DIFFICULTY: ★★☆

IC 434
THE HORSEHEAD

The Horsehead Nebula

IC 434 (The gases behind the horse's head) and Barnard 33 (The dark nebula forming the horse's head) create the famous Horsehead nebula near the Orion nebula. The Horsehead is also right next to NGC 2023 (The small blue reflection nebula visible on the bottom left of the horse) and NGC 2024 (The Flame nebula, visible to the left of the others).

Photographing this group of nebulae is easy, as long as you spend enough time on it. The red gases of IC 434 are faint and will look grainy if the total exposure time is too low. The photo on the right was taken with 4 hours of exposure, using an 8" telescope.

Just like M42, the IC 434 is a perfect target for wide field DSLR astrophotography (left).

COOL FACTS

- Detected in photographs in 1888
- Horse's head would be invisible if there was no colorful gases behind it
- Bright star Alnitak shines light into the flame

DESIGNATION	IC 434 Barnard 33 NGC 2023 NGC 2024
TYPE	Nebula
CONSTELLATION	Orion
MAGNITUDE	6.8
SIZE	8' x 6'

FINDING YOUR TARGET

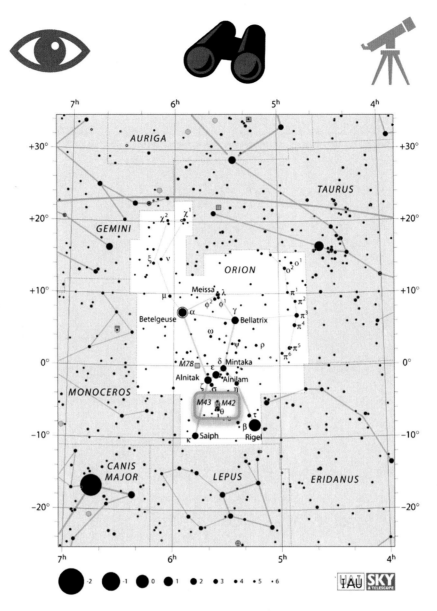

The Horsehead and flame nebula can not be seen with the naked eye, and are too faint to be seen through binoculars. It is a challenging object to find through telescopes, mostly because the horse's head is a dark nebula, while the gas behind it is made of hydrogen. The use of a filter on a large aperture telescope will help spot the group, but both NGC 2023 and NGC 2024 will be more visible than IC 434 and Barnard 33.

The easiest way to find the Horsehead nebula is to locate the bright star Altinak, in Orion's belt. Pointing your telescope at this star will ensure that you are on target, then re-center the nebula before photographing it. Another way to find it, with wide field photography, is to aim your lens towards the Orion nebula.

- Suggested minimum focal length: 50mm
- Ideal focal length: 600mm

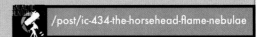
/post/ic-434-the-horsehead-flame-nebulae

WINTER
#11

DIFFICULTY: ★★★

IC 2118
THE WITCH HEAD

The Witch Head Nebula

The Witch Head nebula got its name because it looks like the profile of a witch. It is facing towards the bright blue supergiant star, Rigel.

This target gets a 3-star difficulty because of its faint magnitude of 13, and the fact that processing may be tricky due to its proximity to one of the brightest stars in the sky.

Expect to spend several hours on the Witch or you will only be able to get parts of it. We advise capturing it with a DSLR camera lens (left) before attempting it with a telescope. Because of the faintness of the nebula, it is extremely hard to properly center in the camera without using computer software. The gases will not be revealed on your images until they are stacked.

COOL FACTS

- Discovered with Astrophotography in 1909
- About 900 light-years away from Earth
- Orion's brightest star Rigel shines on the Witch's gases

DESIGNATION	IC 2118
TYPE	Nebula
CONSTELLATION	Eridanus
MAGNITUDE	13
SIZE	3° x 1°

- Suggested minimum focal length: 50mm
- Ideal focal length: 600mm

FINDING YOUR TARGET

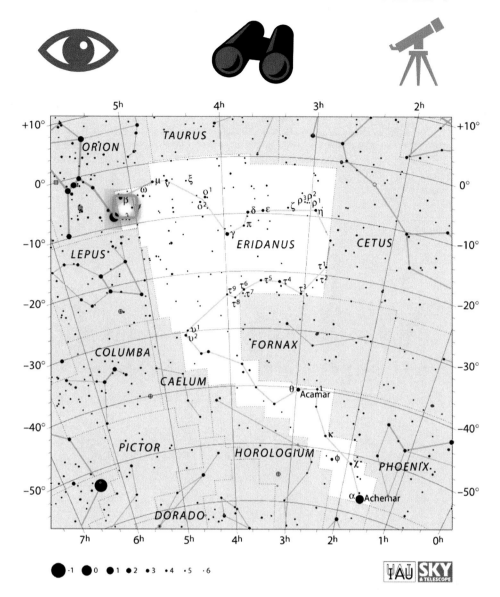

IC 2118 is located just 2.6 degrees east of the 7th brightest star in the night sky: Rigel. Although Rigel is part of Orion, the Witch is instead considered to be in the neighbor constellation of Eridanus.

The Witch Head nebula is faint and difficult to see. It is of course not visible with the naked eye, and is debatable on whether it can be spotted with binoculars or not. Using a pair of wide binoculars under truly dark skies might reveal what seems to be an elongated patch of darkness. We decided to keep the binocular icon red because we were not yet able to see it ourselves after several attempts.

Viewing this nebula with a telescope is also a challenge. The recommended way to do it is to use a small telescope with wide band and narrow filters attached, even then, the Witch Head is far from being impressive visually.

/post/ic2218

WINTER #12

DIFFICULTY: ★★★

MESSIER 78
REFLECTION NEBULA IN ORION

Messier 78

M78 is a diffuse reflection nebula, not far from the Orion nebula, and the Horsehead and Flame nebulae.

Photographing this Messier object is a little bit of a challenge, as you will need long exposure times to capture the details within the dark lanes. Those are passing in front of the two stars that illuminate the nebula.

M78, like its neighbors, is also a great wide field target. With a camera lens, you will be able to capture M78, M42, IC 434, and more. If you would rather get a lot of details in M78 itself, like in the images above, a telescope will be needed. Notice the red gases all around, more information on the next pages.

COOL FACTS

- Discovered in 1780
- Brightest diffuse reflection nebula in the sky
- Has formed about 200 stars

DESIGNATION	M78
TYPE	Nebula
CONSTELLATION	Orion
MAGNITUDE	8.3
SIZE	8' x 6'

- Suggested minimum focal length: 50mm
- Ideal focal length: 700mm

FINDING YOUR TARGET

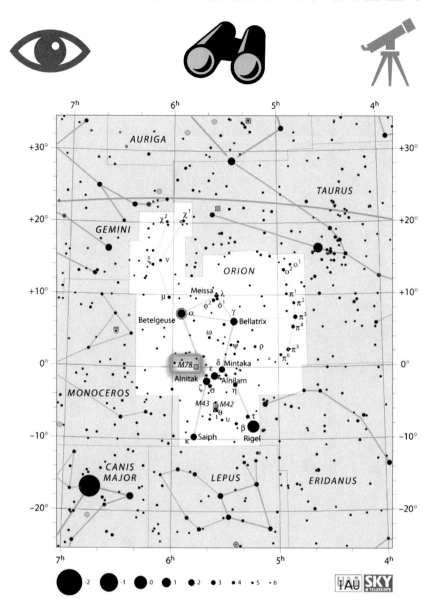

M78 is located in the constellation of the hunter: Orion. The nebula is not visible with the naked eye, but can be seen through small and large binoculars. Most binoculars will show just a faint patch in the sky, but telescopes will reveal the two stars glowing on the nebulosity.

To find M78, first locate Alnitak, in Orion's Belt. Then travel about 2.5 degrees northeast to land on the nebula.

If you are having trouble finding this dim target, attach a lens of about 50mm to your DSLR camera, and do a long exposure of the area around Alnitak. When checking your image, there should be the two bright, but fuzzy stars of M78 in the bottom left of the Flame.

WINTER
#13

DIFFICULTY: ★★★

SH 2-276
BARNARD'S LOOP

Barnard's Loop

Barnard's Loop is an emission nebula that originated from a supernova about 2 million years ago.

The key to photographing this target is to spend as much time as possible getting the red colors from the Loop, as those are really difficult to capture without filters or a modified camera.

This target gets a 3-star difficulty due to the complexity of processing not only the Loop, but all the different nebulae that will be visible in your final image.

The photo on the left is the result of 4 hours of total exposure at 50mm f/1.8, while the one on the right took over 12 hours at 85mm f/4.

COOL FACTS

- Discovered in 1895 with long duration film exposures
- Origin of the Loop is still unknown
- Between 520 and 1440 light-years away

DESIGNATION	Sh 2-276
TYPE	Nebula
CONSTELLATION	Orion
MAGNITUDE	5.0
SIZE	10°

- Suggested minimum focal length: 35mm
- Ideal focal length: 50mm

FINDING YOUR TARGET

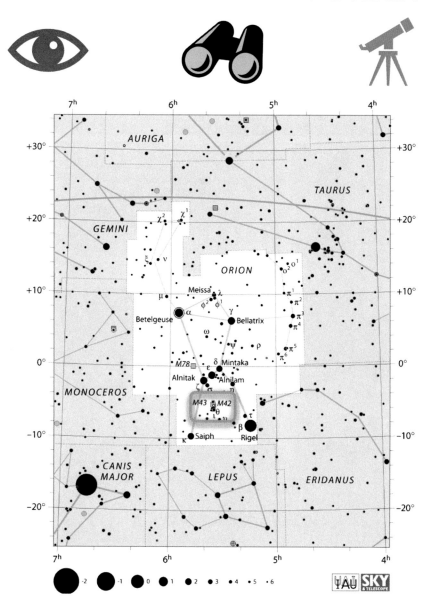

Through photographs, the Loop appears as a bright, red semi-circle with the Orion nebula, and Horsehead and Flame nebulae in its center. From a wide point of view, the nebula looks like a big smile in the sky. Sadly, this is not the case visually.

Barnard's Loop is too faint to be seen with the naked eye. Binoculars and any size telescopes will not reveal much except for a very faint gas in the entire view. It is really difficult to even know if you are looking at the nebula through a telescope because there is no way to compare the gas you see with the actual darkness of space due to its size.

Finding the nebula is extremely easy, it is located in the Orion constellation and you can imagine it just under M42, M78 and IC 434. We suggest spending time with a Hydrogen Alpha filter when capturing this target, or use a modified DSLR/Mirrorless camera if you want the best result possible.

/post/barnards-loop

WINTER #14

DIFFICULTY: ★★☆

NGC 869 & NGC 884
DOUBLE CLUSTER IN PERSEUS

Double Cluster in Perseus

NGC 869 and NGC 884 are two clusters in the constellation of Perseus. Both are very similar to each other in size, magnitude, and age.

When capturing this impressive Double Cluster, you will also be able to image a smaller, third cluster nearby: NGC 957. You may also photograph each cluster individually if using a large telescope.

There is a lot of Hydrogen Alpha gas hidden behind the clusters. Those can be revealed with the use of an HA filter, like we did on the picture on the left. Seeing these stars float in a sea of red gasses is really beautiful, but tricky to process!

COOL FACTS

- Represents the jewels from the handle of Perseus' sword
- About twice the size of the Moon
- Total mass of at least 20,000 solar masses

DESIGNATION	NGC 869 NGC 884
TYPE	Double Cluster
CONSTELLATION	Perseus
MAGNITUDE	NGC 869 - 3.7 NGC 884 - 3.8
SIZE	NGC 869 - 30' NGC 884 - 30'

FINDING YOUR TARGET

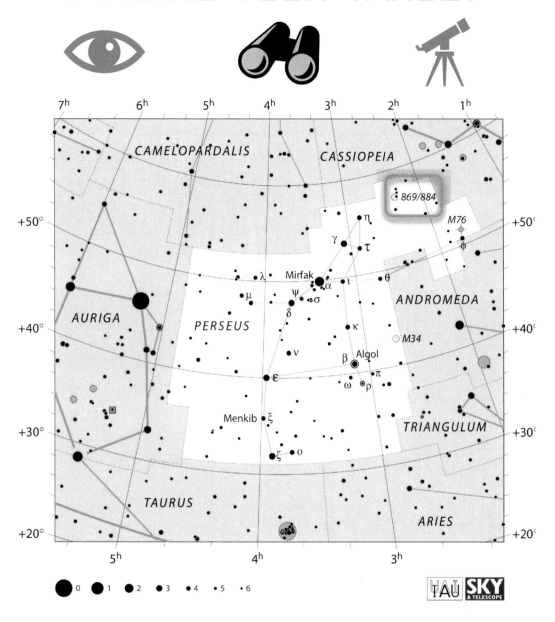

The Double Cluster is located in the constellation of Perseus, but lies just near the "W" shape of Cassiopeia.

Because of their magnitude and size, both NGC 869 and NGC 884 can be seen with the naked eye under clear dark skies, which makes them really easy to find! The Double Cluster can be spotted in the middle of an imaginary line that links the Cassiopeia star Ruchbach, and the highest star in the constellation of Perseus, Miram.

Looking at the double cluster through binoculars is a beautiful sight, with both clusters of stars looking like two worlds dancing together. A telescope will reveal an immense number of faint stars, and a brighter center in each object.

- Suggested minimum focal length: 135mm
- Ideal focal length: 700mm

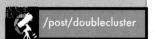

WINTER #15

DIFFICULTY: ★★☆

NGC 2264
THE CHRISTMAS TREE

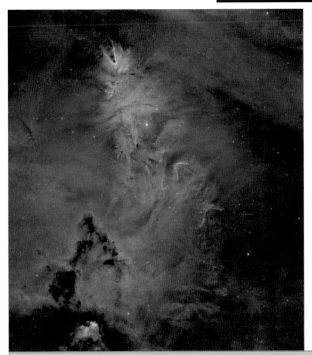

The Christmas Tree Cluster

We end the Winter season with the most festive Deep Sky Object of all! NGC 2264 refers to two objects as one: The Christmas Tree cluster, and the Cone nebula, which forms some of the nebulosity around the cluster. It also contains the Fox Fur nebula and the Snowflake cluster.

This is great for beginners because it is basically four targets in one. The main cluster, as well as the Fox Fur nebula, are easy to photograph and do not require many hours of exposure time. Expect to spend a lot more for the reds to pop out.

The image on the left shows a Narrowband combination (81 hours). The one the right was obtained with a DSLR camera and 4 hours of exposure.

COOL FACTS

- Cluster discovered in 1784, nebulosity found 2 years later the day after Christmas
- Lies in the Orion arm of the Milky Way
- Located 2,600 light-years away

DESIGNATION	NGC 2264
TYPE	Nebula
CONSTELLATION	Monoceros
MAGNITUDE	3.9
SIZE	20'

- Suggested minimum focal length: 85mm
- Ideal focal length: 650mm

FINDING YOUR TARGET

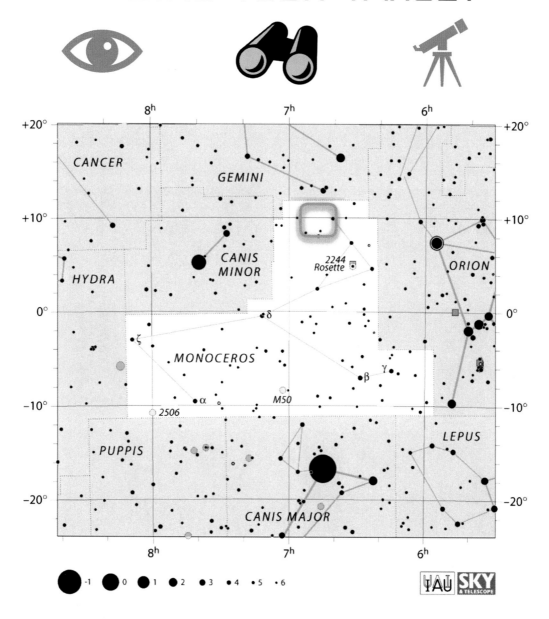

The Christmas Tree cluster and its nebulosity are located in the constellation of Monoceros, not far from the Rosette nebula.

Finding the tree is not easy, as the stars in Monoceros are faint, but can be found with less hassle by star hopping from the surrounding constellations. Start from the famous shoulder of Orion: Betelgeuse, or Beetle Juice. Then, jump to Alhena, which forms one of the feet of Pollux in the constellation of Gemini, and hop to another star, just next to Alhena, called Xi Gem. The Christmas tree can be found just 3 degrees south of that star.

The Christmas Tree cluster can be seen with the unaided eye and is easily spotted with binoculars. It becomes a stunning target through any size telescope, as you can easily resolve the shape of a Christmas tree. The Cone nebula and the gases behind the cluster are much more difficult to see, but you might be able to see it in large telescopes, under very dark skies.

/post/ngc2264

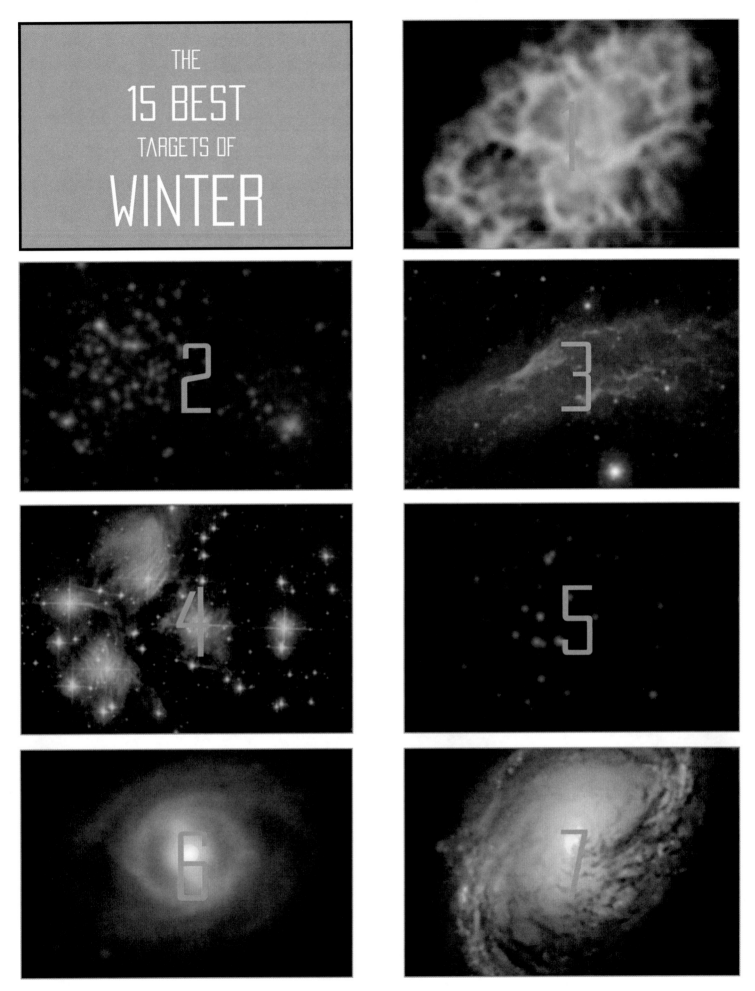

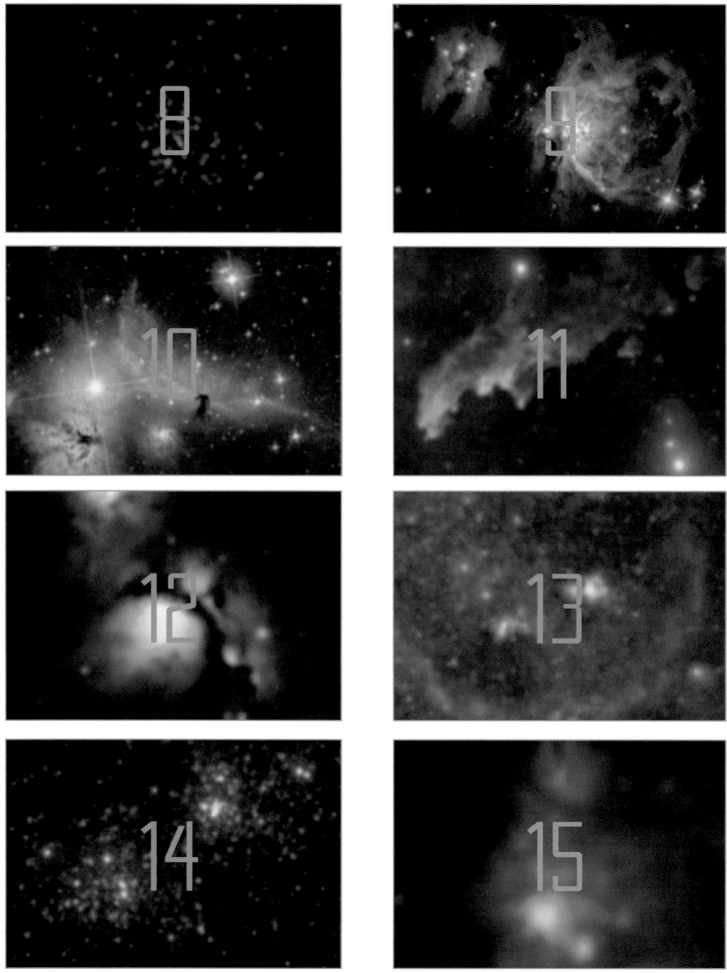

SUMMARY

NUMBER	IMAGE	DESIGNATION	MAGNITUDE	SIZE	DIFFICULTY
1		VIRGO CLUSTER	9.0 - 15.0	1° +	★★★
2		M104	8.98	8'.7 X 3'.5	★★☆
3		NGC 2237	9.0	1.3°	★☆☆
4		M101	7.86	28'.8 X 26'.9	★★☆
5		M66 GROUP	8.9 - 10.25	VARIES	★★★
6		M63	9.3	12'.6 X 7'.2	★★☆
7		M81 & M82	6.94 & 8.41	26'.9 X 14'.1 & 11'.2 X 4'.3	★★☆
8		M97	9.9	3'.4 X 3'.3	★★★
9		NGC 4631	9.8	15'.5 X 2'.7	★★★
10		M51	8.4	11'.2 X 6'.9	★★☆
11		IC 443	12.0	50'	★★★
12		NGC 4565	9.5	15'.90 X 1'.85	★★★
13		M64	9.36	10'.71 X 5'.13	★★★
14		M106	9.1	18'.6 X 7'.2	★★☆
15		NGC 2359	11.45'	8' X 8'	★★★

NUMBER	IMAGE	DESIGNATION	MAGNITUDE	SIZE	DIFFICULTY
16		M8	6.0	90' X 40'	★☆☆
17		NGC 6888	7.4	18' X 12'	★★☆
18		M20	6.3	28'	★☆☆
19		M27	7.5	8' X 5'.6	★☆☆
20		M17	6.0	11'	★★☆
21		M57	8.8	230" X 230"	★★★
22		M24	4.6	90'	★★☆
23		M16	6.0	30'	★★☆
24		RHO OPHIUCHI	4.6	4.5° X 6.5°	★★★
25		NGC 7000 & NGC 5070	4.0 & 8.0	120' X 100 & 60 X 50'	★★★
26		M11	6.3	14'	★★☆
27		NGC 7380	7.2	25'	★★★
28		M13	5.8	20'	★★☆
29		NGC 6960	7.0	3°	★★☆
30		M75	9.18	6'.8	★★★

NUMBER	IMAGE	DESIGNATION	MAGNITUDE	SIZE	DIFFICULTY
31		M31	3.44	178' X 63'	★☆☆
32		M15	6.2	18'	★★☆
33		IC 1805	18.3	150' X 150'	★★☆
34		IC 1848	6.5	150' X 75'	★★☆
35		M33	5.72	73' X 45'	★☆☆
36		NGC 7293	7.6	25'	★☆☆
37		IC 1396	3.5	170' X 140'	★★★
38		HCG 92	14.0	3.5	★★★
39		NGC 7635	10.0	16' X 9'	★★☆
40		NGC 7023	6.8	18' X 18'	★★★
41		IC 5146	7.2	12'	★★★
42		M74	10.0	10'.5 X 9'.5	★★☆
43		M92	6.3	14'	★★☆
44		IC 405	6.0	37' X 19'	★★☆
45		NGC 281	7.0	20' X 30'	★☆☆

NUMBER	IMAGE	DESIGNATION	MAGNITUDE	SIZE	DIFFICULTY
46		M1	8.4	420" X 290"	★★☆
47		M35	5.3	28'	★★☆
48		NGC 1499	6.0	2.5°	★☆☆
49		M45	1.6	110'	★☆☆
50		M44	3.7	95'	★★☆
51		M95	11.4	3'.1 X 2'.9	★★☆
52		M96	10.1	7'.6 X 5'.2	★★☆
53		M37	6.2	24'	★★☆
54		M42	4.0	65' X 60'	★☆☆
55		IC 434	6.8	8' X 6'	★★☆
56		IC 2118	13	3° X 1°	★★★
57		M78	8.3	8' X 6'	★★★
58		BARNARD'S LOOP	5.0	10°	★★★
59		NGC 869 & NGC 884	3.7 & 3.8	30' & 30'	★★☆
60		NGC 2264	3.9	20'	★★☆

END WORD

It is our hope that this guide will be useful to you during your time exploring deep sky objects in the night sky. Be advised that all targets listed were chosen because they are, in our opinion, the best suited for new astrophotography hobbyists.

Just like the universe, expand your interest and try targets that we did not mention in this book, especially if you happen to upgrade your equipment to a better camera or a bigger telescope!

For a complete gallery of our images in high definition, our tutorials, and to follow our adventures in the Nevada desert under thousands of stars, find us at galactic-hunter.com and on youtube.com/galactichunter.

Clear Skies,

Galactic Hunter

CREDITS

Antoine & Dalia Grelin would like to thank all the following people who contributed to the beautiful illustrations of this book.

All constellation maps used in this book were created by IAU and Sky & Telescope magazine (Roger Sinnott & Rick Fienberg)

Example #1: 2/3: Pages 2 and 3, full size image
Example #2: 146-3: Page #146, Third Image.

2/3 Galactic Hunter; **4/5** ESA/Hubble & NASA; **6/7** ESA/Hubble & NASA; **8/9** Galactic Hunter; **10/11** NASA, ESA, S. Beckwith (STScI), and the Hubble Heritage Team (STScI/AURA); **12-1** ESO; **12-2** Connor Matherne, Deep Sky West; **14-1** Galactic Hunter; **14-2** NASA and The Hubble Heritage Team (STScI/AURA) **16-1** Trevor Jones; **16-2** Galactic Hunter; **18-1** Galactic Hunter; **18-2** NASA, ESA, CXC, SSC, and STScI; **20-1** NASA, ESA and the Hubble Heritage (STScI/AURA); **20-2** Pavle Gartner; **22-1** Alexandre Steen; **22-2** ESA/Hubble & NASA; **24-1** NASA, ESA and the Hubble Heritage (STScI/AURA); **24-2** Galactic Hunter; **26-1** Neil Winston; **26-2** Lord Rosse; **28-1** Connor Matherne, Deep Sky West; **28-2** NASA & ESA; **30-1** Galactic Hunter; **30-2** NASA, ESA, S. Beckwith (STScI), and the Hubble Heritage Team (STScI/AURA); **32-1** Gibran Pierluissi; **32-2** Nicolas Kizilian; **34-1** Jérémie Bottollier Curtet; **34-2** Nicolas Kizilian; **36-1** Jérémie Bottollier Curtet; **36-2** NASA and The Hubble Heritage Team (AURA/STScI); **38-1** Pavle Gartner; **38-2** NASA, ESA, the Hubble Heritage Team (STScI/AURA); **40-1** ESO/B. Bailleul; **40-2** ESO/Digitized Sky Survey 2. Acknowledgement: Davide De Martin; **44-45** Hubble Heritage Team (AURA/STScI/NASA); **46-1** Connor Matherne, Deep Sky West; **46-2** Galactic Hunter; **48-1** Nicolas Kizilian; **48-2** Neil Winston; **50-1** ESO; **50-2** Galactic Hunter; **52-1** Jérémie Bottollier Curtet; **52-2** Jean-François Godin; **54-1** Neil Winston; **54-2** Connor Matherne, Deep Sky West; **56-1** Hubble Heritage Team (AURA/STScI/NASA); **56-2** Galactic Hunter; **58-1** Connor Matherne; **58-2** Neil Winston; **60-1** NASA and ESA; **60-2** Galactic Hunter; **62-1** Ryan Jones; **62-2** Connor Matherne; **64-1** Ryan Jones; **64-2** Galactic Hunter **66-1** NASA, ESA, STScI and P. Dobbie; **66-2** Galactic Hunter; **68-1** Alexandre Steen; **68-2** Connor Matherne, Deep Sky West; **70-1** NASA, ESA, and the Hubble Heritage Team (STScI/AURA); **70-2** Galactic Hunter; **72-1** Galactic Hunter; **72-2** Chris Sanford; **74-1** NASA and The Hubble Heritage Team (STScI/AURA); **74-2** NASA and The Hubble Heritage Team (STScI/AURA); **78-79** Galactic Hunter; **80-1** Galactic Hunter; **80-2** Galactic Hunter; **82-1** NASA, ESA; **82-2** NASA, ESA; **84-1** Andrew Klinger; **84-2** Connor Matherne, Deep Sky West; **86-1** Galactic Hunter; **86-2** Adam Butko; **88-1** Galactic Hunter; **88-2** Galactic Hunter; **90-1** NASA, ESA, C.R. O'Dell (Vanderbilt University), and M. Meixner, P. McCullough, and G. Bacon (Space Telescope Science Institute); **90-2** Galactic Hunter; **92-1** Ryan Jones; **92-2** Connor Matherne, Deep Sky West; **94-1** NASA, ESA, and the Hubble SM4 ERO Team; **94-2** Jérémie Bottollier Curtet; **96-1** NASA, ESA, Hubble Heritage Team; **96-2** Ryan Jones; **98-1** Adam Butko; **98-2** Nicolas Kizilian; **100-1** Galactic Hunter; **100-2** Connor Matherne, Deep Sky West; **102-1** NASA, ESA and the Hubble Heritage (STScI/AURA)-ESA/Hubble Collaboration; **102-2** NASA, ESA and the Hubble Heritage (STScI/AURA)-ESA/Hubble Collaboration; **104-1** ESA/Hubble & NASA; **104-2** ESA/Hubble & NASA; **106-1** Galactic Hunter; **106-2** Nicolas Kizilian; **108-1** Donovan Campbell-Gillies; **108-2** Ryan Jones; **112-113** Galactic Hunter; **114-1** Galactic Hunter; **114-2** NASA, ESA, J. Hester and A. Loll; **116-1** ESO/S. Brunier; **116-2** Alexandre Steen; **118-1** Galactic Hunter; **118-2** Connor Matherne, Deep Sky West; **120-1** Galactic Hunter; **120-2** Galactic Hunter; **122-1** Alexandre Steen; **122-2** Jérémie Bottollier Curtet; **124-1** Galactic Hunter; **124-2** ESO; **126-1** ESA/Hubble & NASA and the LEGUS Team; **126-2** Galactic Hunter; **128-1** Galactic Hunter; **128-2** Galactic Hunter; **130-1** Antoine Mangiavacca; **130-2** Galactic Hunter; **132-1** Antoine Mangiavacca; **132-2** Galactic Hunter; **134-1** Galactic Hunter; **134-2** NASA/STScI Digitized Sky Survey/Noel Carboni; **136-1** ESO/Igor Chekalin; **136-2** Connor Matherne, Deep Sky West; **138-1** Galactic Hunter; **138-2** Antoine Mangiavacca; **140-1** Galactic Hunter; **140-2** Orlando DeJesus; **142-1** Galactic Hunter; **142-2** Alexandre Steen; **152** Galactic Hunter; **153** Galactic Hunter

Made in the USA
Middletown, DE
15 July 2024